STRUCTURES AND SOLID BODY MECHANICS SERIES

General Editor B. G. NEAL

Braced Frameworks

braced frameworks

An Introduction to the Theory of Structures

SECOND EDITION

E. W. Parkes
Professor of Mechanics in the University of Cambridge,
formerly Professor of Engineering, Leicester University.

PERGAMON PRESS
Oxford · New York · Toronto · Sydney

Pergamon Press Ltd., Headington Hill Hall, Oxford
Pergamon Press Inc., Maxwell House, Fairview Park, Elmsford,
New York 10523
Pergamon of Canada Ltd., 207 Queen's Quay West, Toronto 1
Pergamon Press (Aust.) Pty. Ltd., 19a Boundary Street,
Rushcutters Bay, N.S.W. 2011, Australia

Copyright © 1965 Pergamon Press Ltd.

All Rights Reserved. No part of this publication may be reproduced, stored in a retrieval system, or transmitted, in any form or by any means, electronic, mechanical, photocopying, recording or otherwise, without the prior permission of Pergamon Press Ltd.

First edition 1965
Second edition 1974

Library of Congress Cataloging in Publication Data
Parkes, Edward Walter, 1926–
Braced frameworks.
(Structures and solid body mechanics series)
Bibliography: p.
1. Structures, Theory of. I. Title.
TA645.P3 1975   624'.17   74-10556
ISBN 0 08 018078 7
ISBN 0 08 018077 9 (pbk.)

Printed in Great Britain by Biddles Ltd, Guildford, Surrey

contents

PREFACE ix

1 BASIC CONCEPTS	1.1	Function and failure	1
	1.2	Analysis and synthesis	3
	1.3	Equilibrium	3
	1.4	Compatibility	4
	1.5	Environment and deformation	6
	1.6	Statical determinacy	14
	1.7	Methods of structural analysis	17
	1.8	Braced and non-braced frames	22
		Examples	24
2 STATICALLY DETERMINATE TRUSSES	2.1	Introduction	27
	2.2	Pin-joints and pin-ended members	27
	2.3	Equilibrium at joints and supports	29
	2.4	The forces in the members of a truss	31
		2.4.1 Successive resolution at the joints	
		2.4.2 Polygons of forces at the joints	
		2.4.3 Maxwell diagrams and Bow's notation	
		2.4.4 Method of sections	
		2.4.5 Henneberg's method	
		2.4.6 Tension coefficients	

v

CONTENTS

	2.5 Conditions for statical determinacy	44
	2.6 Superposition of loads on a truss	48
	2.7 Influence lines	50
	2.8 The displacements of trusses	55
	2.8.1 Displacement diagrams	
	2.8.2 Algebraic solution	
	2.8.3 Virtual work	
	2.9 Hooke's law and the superposition of displacements	68
	2.10 Properties of symmetrical structures	69
	2.11 Maxwell's reciprocal theorem	77
	Examples	82
3 REDUNDANT TRUSSES	3.1 Introduction	85
	3.2 Number and choice of redundancies	87
	3.3 Direct comparison of lengths	89
	3.4 Energy methods	100
	3.4.1 The energies of a structure	
	3.4.2 Energy theorems	
	3.4.3 The energies of linear-elastic bars	
	3.5 Solution by energy methods	110
	3.6 Relaxation methods	121
	Examples	127
4 SECONDARY STRESSES AND THE FAILURE OF BRACED FRAMEWORKS	4.1 Analytical assumptions	129
	4.2 Members subjected to end load and bending	130
	4.3 Solution by slope-deflection equations	134
	4.4 Solution by moment distribution	137
	4.5 The failure of braced frameworks	141
	4.5.1 Types of failure	
	4.5.2 Criteria of safety	
	4.5.3 Failure of an individual member	
	4.5.4 Failure of a joint	
	4.5.5 Failure of a complete structure	
	Examples	158

CONTENTS

5 MINIMUM WEIGHT FRAMEWORKS	5.1 Maxwell's lemma	160
	5.2 Michell's theorem	165
	5.3 The form of the optimum structure	168
	5.4 Michell frameworks	175
	5.5 The symmetrical three-force system	181
	5.6 Graphical methods	184
	Examples	189

ANSWERS TO EXAMPLES   191

REFERENCES   193

INDEX   195

preface

BRACED frameworks are a particular type of construction which has application in almost every branch of engineering. They are found in bridges, in cranes, in aircraft, in electrical transmission towers, in roofs and in astronomical telescopes. The Red Indian builds his wigwam and the farmer makes a gate on a braced framework.

Braced frameworks are, however, only one of several different kinds of structural system, and the theory of structural systems is itself only one aspect of the study of engineering science, which is again but a part of the general subject of engineering, which includes not only science, but a good deal of arts and social science as well. It may therefore seem strange, in a book aimed mainly at the student, to limit oneself to a particular branch of structural theory. It is important for the reader to realise that although the examples used in this book are limited to braced frameworks, the ideas which are developed are applicable to all kinds of structural system and, indeed, to many other branches of engineering as well. The student who can understand the methods of analysis and synthesis described here should have no difficulty in extending them to other types of construction such as beams, plates and shells.

The author has chosen to write a book on "Braced Frameworks" for three main reasons. First, the simplicity of this form of

PREFACE

construction enables one to demonstrate the general ideas of structural analysis and synthesis in ways most readily understandable by the beginner. Second, recent advances in the theory of non-braced frames have produced a spate of excellent books dealing with this type of structural system, but the braced framework has consequently been rather neglected. Third, it has been his experience as an engineering consultant that many of those concerned with the design of braced frameworks might welcome a book which treats the subject more fully than is possible in the one or two chapters usually devoted to it in a general structures textbook.

It is too much to hope that the first edition of a new book of this kind will be entirely free from error, and the author will be most grateful to his readers for their help in remedying such defects as may exist.

preface to the second edition

THE principal differences from the first edition are the change to S.I. units and the addition of references to recent work.

1. basic concepts

1.1. Function and Failure

This book is mainly concerned with a particular form of construction—that known as the braced framework—but the general principles and methods of analysis discussed in this first chapter are applicable to any type of structural system, and the reader who can understand the arguments used in terms of the braced framework should have little difficulty in applying his knowledge to such structures as beams, plates and shells.

In analysing an existing structure, or designing a new one, the first piece of information which we require is a specification of the function which the structure has to fulfil. If the structure is a building, it will have to carry certain loads on each floor, together with external loads applied to the roof and walls by wind or snow. If the structure is a railway bridge it will have to carry dynamic loads due to trains moving across it. If the structure is a gas turbine blade it will be subjected to dynamic loads due to rotation and gas flow, together with sudden heating and cooling. If the structure is a wooden window frame it will be subjected to changes of humidity.

All of these changes in the environment (changes in load, temperature, humidity, etc.) produce dimensional changes in the structure. These dimensional changes may or may not be recovered if the environment is restored to its original condition. A structure is said to be satisfactory if the deformations due to performing its desired function are less than certain prescribed limits. No general limits can be set for permitted deformations, since these vary both with the structure and its function. A deflection which would be

(a) Initial shape

(b) Deformation acceptable

(c) Deformation not acceptable (failure)

(d) Deformation not acceptable (failure)

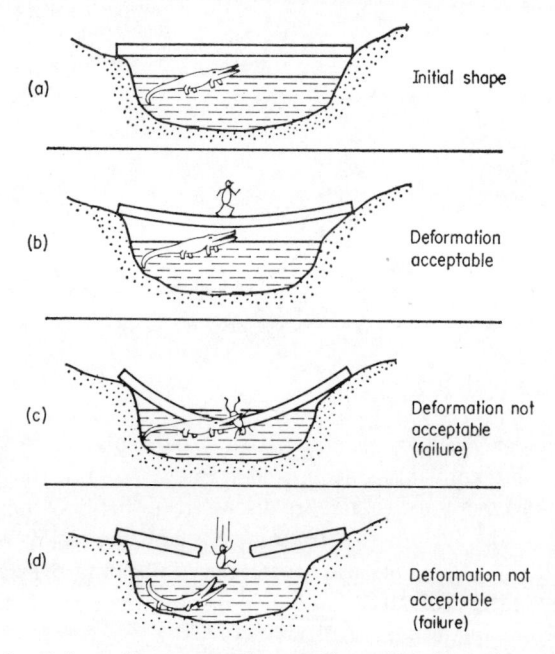

FIG. 1. Failure of a structure (from *Bulletin of Mechanical Engineering Education*, July 1959)

acceptable for the wing of an aeroplane would be much too large if it occurred in a bridge of the same span, and a deflection which would be acceptable for the bridge would again be much too large if it occurred in a telescope mounting.

If the deformations of the structure exceed the limits set, the structure is said to have failed. It is important to realise that it is not necessary for the structure to break for failure to occur. This is illustrated in Fig. 1, where the difference between condition (c) and (d) is of no significance to the traveller.

1.2. Analysis and Synthesis

The determination of whether a given structure will satisfactorily fulfil a given function or set of functions is known as the analysis of the structure: the determination of the best possible structure to fulfil a given function or set of functions is known as the synthesis (or design) of the structure.

The design of a structure usually depends partly upon an accumulated experience of analysis and partly upon economic, sociological and aesthetic factors which are outside the scope of the present book. Only in the simplest cases is it possible to proceed directly to the design of the optimum structure to do a given job. In Chapter 5 we show for a particular type of optimisation (minimum weight) how direct design of a framework may be possible. The more usual method of design is indirect: the designer guesses a structure, analyses it to see whether it is satisfactory, and improves his guess by using the results of this analysis. The process may be repeated a number of times.

The great importance of structural analysis lies in its use in the indirect method of design, and it is with the structural analysis of frameworks that the majority of this book is concerned. Except under the very limited conditions of Chapter 5, we shall make no attempt to discuss structural synthesis.

1.3. Equilibrium

Every structure, and every part of every structure, must be in equilibrium under the forces applied to it. The equilibrium may be static or dynamic. In dynamic problems it is often convenient to reduce the system to an equivalent set of static forces by using d'Alembert's principle.

Figure 2(a) shows a bridge truss in equilibrium under the loads applied to it and the reactive forces at the supports (for simplicity the weight of the bridge itself is ignored). Figure 2(b) shows part of the truss in equilibrium under the load applied to it and the reactive forces from the support and the rest of the structure. Figure 2(c) shows one joint of the truss in equilibrium.

BRACED FRAMEWORKS

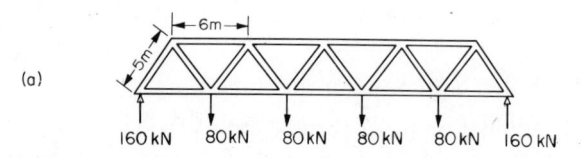

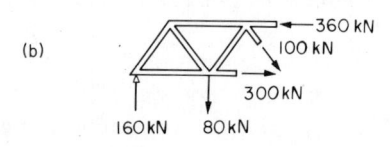

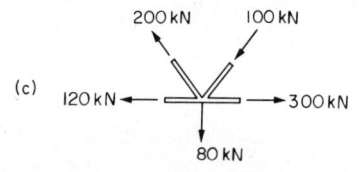

Fig. 2. Equilibrium of a structure, and of parts of a structure

A clear understanding of the principle of equilibrium is of the greatest importance in structural analysis, and the student should take every opportunity to consider the forces acting on complete structures and on parts of structures, and convince himself that they form a system in equilibrium.

1.4. Compatibility

The second great principle of structural analysis is that of geometrical compatibility. The parts of a structure must fit together.

Consider the roof truss shown in Fig. 3(a), supported by three stanchions. Suppose that by some means we calculated the deformed shape of the roof truss under load and that our result appeared as shown in Fig. 3(b). Then, clearly, our calculations were incorrect, since this solution violates the compatibility condition that the roof truss rests on the centre stanchion. The true solution must be similar to that shown in Fig. 3(c).

Ensuring correct fit, or compatibility of deformation, is of fundamental significance in all kinds of structural analysis, and it is very

BASIC CONCEPTS

important that the student should understand this principle and should acquire skill in applying it. In many of the more sophisticated methods of structural analysis (e.g. those described in §3.5 of this book), the compatibility condition is disguised as an energy criterion. There is a grave danger that the student may learn these energy

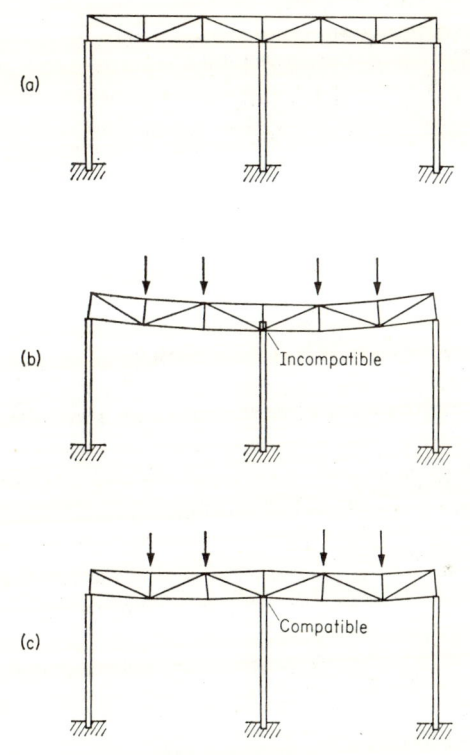

FIG. 3. Geometrical compatibility

methods by rote without properly understanding the geometrical conditions upon which they are based: under these circumstances he not infrequently misapplies the energy methods and obtains incorrect solutions. Whenever any doubt arises as to the validity or correctness of one of the sophisticated methods of analysis, it can always be resolved by reverting to fundamental concepts and considering the geometrical requirement that the structure must fit together.

1.5. Environment and Deformation

The principles of equilibrium and compatibility are sufficient in themselves for the solution of certain types of structural problems, but generally one further matter has to be considered—the characteristics of the material of construction. Braced frameworks (the nominal subject of the present book) are structures composed of bars (more or less straight) fastened together at their ends. To simplify our present discussion, then, let us consider a straight bar of our material, of length l_0 and uniform cross-sectional area A_0, immediately after manufacture (Fig. 4). The bar will at this time

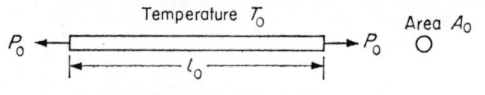

FIG. 4. Bar in datum state at time t_0

(t_0) be in an environment characterised by the values of certain physical quantities. The most important of these for our present purposes are the temperature (T_0) and the end load applied to the bar (P_0). Other quantities which may be important are the humidity and the intensity of radiation.

If one or more of the parameters defining the environment of the bar in the datum state (t_0, T_0, P_0, etc.) change to new values (t, T, P, etc.) the dimensions of the bar may change. A knowledge of the relationships between changes of environment and the deformation of structural elements is essential to most forms of structural analysis. We give some examples below.

Figure 5(a) shows the increase in length of an 0·4 per cent carbon steel bar where the only environmental parameter which is changing is time. The bar is initially 100cm long and it has an initial cross-sectional area of 8cm². The temperature and end load remain constant at their datum values of 450°C and 200kN. It will be seen that after 1000hr the length of the bar has increased to 100·9cm. Dimensional changes which occur due to the passage of time at constant temperature and load are said to be due to "creep".

Figure 5(b) shows the increase in length of a copper bar due to change of temperature. The bar is unloaded and at 0°C it has a

BASIC CONCEPTS

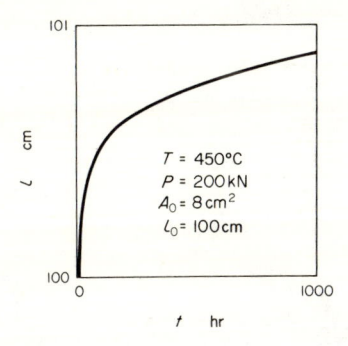

FIG. 5(a). Creep of 0·4% carbon steel bar

length of 20cm. It will be seen that at 1000°C the length has increased to 20·4cm.

Figure 5(c) shows the increase in length of an aluminium alloy bar as the end load applied to it is increased (it is assumed that the load is applied in such a way that the bar extends and does not bend). The bar is initially of length 100cm and it has an initial cross-sectional area of 20cm². The temperature remains constant at its datum value of 20°C. It will be seen that under a load of 800kN the length of the bar increases to 101cm.

It is important to realise that restoring the environmental conditions to the datum values does not necessarily ensure that the bar will return to its original dimensions. In the example of Fig. 5(a), return to the datum conditions is not possible, since the varying parameter is time.

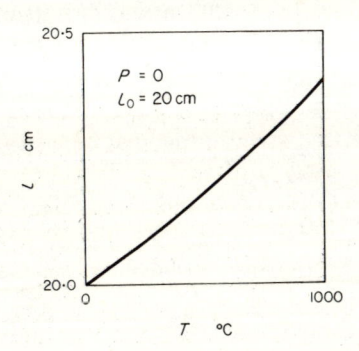

FIG. 5(b). Expansion of copper bar

BRACED FRAMEWORKS

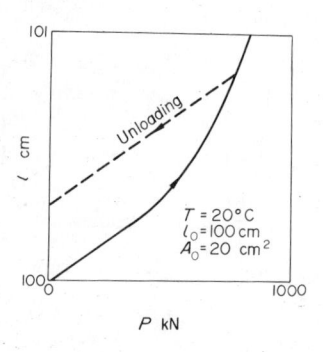

FIG. 5(c). Extension of aluminium alloy bar

In Fig. 5(b), restoring the temperature to 0°C will cause the copper bar to contract to its original length of 20cm. In Fig. 5(c), removing the load will not generally restore the bar to its original length—the loading and unloading paths are not the same. The dotted line in Fig. 5(c) shows the effect of unloading from a maximum load of 750kN. It will be seen that the final length of the bar is 100·3cm. The bar is said to have a "residual extension" or "permanent set".

A further important point is that the effects we have been describing are not generally additive. The increase in length of a bar due to a temperature rise of 300°C combined with a load of 400kN cannot usually be obtained by adding together the changes in length produced by a temperature rise of 300°C and a load of 400kN, applied separately. The effects of the environmental changes interact.

The relations between the dimensions of a structural member and the environment to which it is subjected are essentially empirical and generally complicated, but it is usually necessary for the purposes of structural analysis to replace the experimental results with simple laws which make the structural system amenable to mathematical treatment. The validity of these laws and the results which follow from them depends entirely on how closely the laws approximate to the experimental environment–deformation relationships. The skill of the structural analyst lies in choosing laws which represent the experimental results sufficiently well, without making the subsequent mathematics too complicated.

One set of environment–deformation relationships in common use is the concept of a "linear system". Let us suppose that the deform-

BASIC CONCEPTS

ations of a structural element are proportional to the changes in the environment which cause them. Typical relationships which satisfy this requirement are shown in equations (1) and (2), where a, b and c are constants.

$$l - l_0 = a(P - P_0) + b(T - T_0) \qquad (1)$$

$$l - l_0 = a(P - P_0) + c \int_{t_0}^{t} P \, dt \qquad (2)$$

In equation (1), the change of length is composed of two terms, one proportional to change of load and the other to change of temperature: time does not enter into the equation and it is assumed that creep does not occur. The effects of load and temperature are separate and additive.

Equation (2) is only applicable at a particular temperature. The first term is similar to that in equation (1). The second (creep) term depends on the integral of the load with respect to time. At constant time, change in length is proportional to load; at constant load, change in length is proportional to time; but the effects of load and time are not additive, since they interact in the second term.

A system in which the deformations are proportional to the environmental changes and in which the effects of the environmental changes do not interact will be described as a "linear system". Structures in which deformations are proportional to changes in individual environmental parameters (all the others being held constant), but in which the environmental effects interact, are not truly linear systems and must be treated with caution. Equation (1) represents a linear system provided a and b are really constants (in particular, a must not be temperature-dependent, nor b load-dependent).

A linear system is usually the most convenient mathematically, but other types of relation must be used where the linear law is not a sufficiently good approximation to the experimental data. Some structural materials (in particular, mild steel) behave linearly under load up to a certain intensity, and thereafter the extension increases greatly at nearly constant load—a typical graph is shown in Fig. 6(a). A material whose deformation is proportional to the applied load is said to be "linear elastic" (commonly abbreviated to "elastic", although the term strictly refers to a material in which the loading and unloading curves are identical). A material whose deformation increases greatly at constant load is said to be "plastic". A material which exhibits both properties, such as mild steel, is said to be

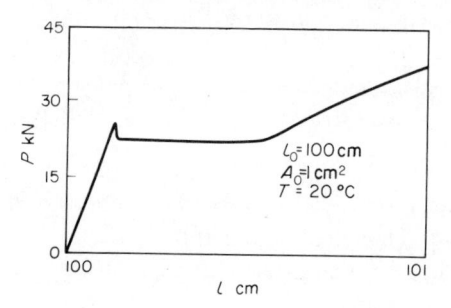

FIG. 6(a). Extension of mild steel bar

"elasto-plastic". An idealised load-deformation curve for a bar made from an elasto-plastic material is shown in Fig. 6(b). Sometimes the elastic deformations can be neglected compared with the plastic ones, and it is then convenient to regard the material as rigid-plastic (Fig. 6(c)).

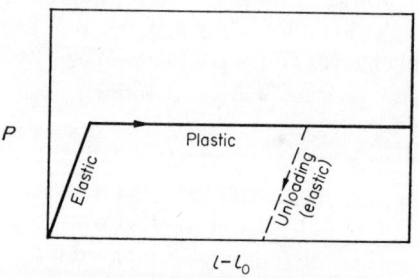

FIG. 6(b). Load–extension curve for elasto-plastic bar

As well as the analytical simplication attendant on replacing the empirical data with simple laws, a further important piece of codification is possible with environment–deformation relationships. It would clearly be very difficult if it was necessary to test every possible size and shape of bar, made of every possible material, which might be built into a braced frame, in order to find its deformation under various environmental changes. Fortunately this is not necessary. Certain well-established experimental relationships enable

BASIC CONCEPTS

us to interpret the results of tests on one bar to another bar made of the same material but geometrically different.

We begin by considering temperature effects. It is found experimentally that in a structure made of one homogeneous isotropic material (i.e. a material whose properties do not vary with position or direction), the increase in distance between two points due to a uniform rise in temperature is proportional to the original distance between them. It follows that if we divide the change of distance by

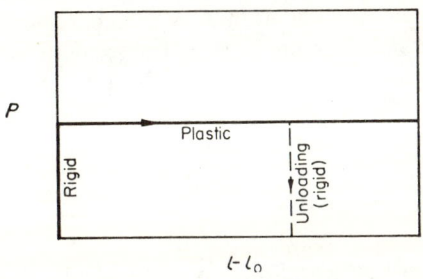

FIG. 6(c). Load–extension curve for rigid-plastic bar

the original distance we obtain a quantity which is constant throughout the body. This quantity is known as the "thermal strain". It may be regarded as the increase in linear dimension per unit distance. Since the thermal strain depends only upon the material of construction and the temperature, experimental data for the change in length of any bar due to temperature rise can clearly be used to predict the change in length of any other bar made of the same material. In a linear system the thermal strain is proportional to the temperature rise. It follows that we may write

$$\text{thermal strain} = \frac{l-l_0}{l_0} = \alpha(T-T_0) \qquad (3)$$

where α is a constant known as the "coefficient of thermal expansion". Since the thermal strain is dimensionless, α has the units of (temperature)$^{-1}$. Typical values are given in Table 1 below.

Experimental data for the behaviour of a bar under end load can similarly be generalised. Suppose we have the load–extension curve for a bar of original length l_0 and original cross-sectional area A_0. Then it is found that if we take a bar of the same material and the same length, but of cross-sectional area $2A_0$, the same extensions

BRACED FRAMEWORKS

are obtained at each stage provided the applied load is doubled. In other words, the important parameter, if we are to generalise our results, is not the load on the bar but the load divided by the original cross-sectional area. This quantity, which may be regarded as the intensity of force per unit area, is known as the "direct stress" and is given the symbol σ. We thus have

$$\text{direct stress} = \sigma = \frac{P}{A_0} \tag{4}$$

The quantity given by equation (4), sometimes known as "engineers'" stress, is the one commonly used in calculations. When a bar is pulled, however, as well as l increasing the cross-sectional area A decreases. In some applications it is important to distinguish between the direct stress as defined by equation (4) and the "true direct stress" (P/A). In our subsequent work we shall mean by "direct stress" P/A_0.

As in the case of temperature effects, it is found experimentally that the increase in length of a bar of constant material and cross-section subjected to end load is proportional to the length of the bar. It follows that we may conveniently define a "direct strain", or increase of length per unit length (given the symbol ε), as

$$\text{direct strain} = \varepsilon = \frac{l - l_0}{l_0} \tag{5}$$

and that we may expect a relationship between σ and ε, determined for one bar, to be applicable to all bars made from the same material and at the same temperature.

In some applications a different definition of direct strain is used. If we regard the increase in strain due to a small change (δl) in the length of the bar as $\delta l/l$, where l is the length at that time, then we obtain the "true direct strain" by integration as

$$\text{true direct strain} = \int_{l_0}^{l} \frac{\delta l}{l} = \ln \frac{l}{l_0} = \ln \left(1 + \frac{l - l_0}{l_0} \right)$$

$$= \frac{l - l_0}{l_0} - \frac{1}{2} \left(\frac{l - l_0}{l_0} \right)^2 + \frac{1}{3} \left(\frac{l - l_0}{l_0} \right)^3 \ldots \tag{6}$$

By comparing equations (5) and (6) it will be seen that the two definitions of strain are similar if the strains are small. In our

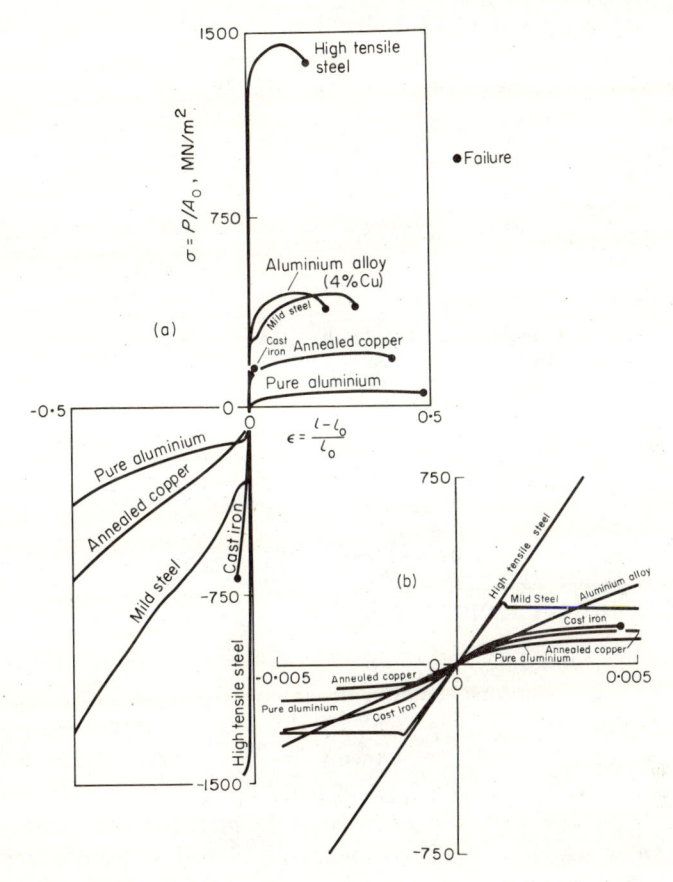

FIG. 7. Stress–strain curves for various materials. (*a*) Complete curves. (*b*) Part of (*a*) reproduced with the strain scale multiplied by 100

subsequent work we shall mean by "direct strain" that defined by equation (5).

The relationship between σ and ε, known as the stress–strain curve, is one of the most important structural properties of materials: examples are given in Fig. 7. If the experimental stress–strain curve may be approximated by a linear law, strain is proportional to stress and we write

$$\varepsilon = \frac{\sigma}{E} \qquad (7)$$

BRACED FRAMEWORKS

where E is a constant known as Young's modulus (Young, 1807). The value of E in compression (when the bar is pushed) is generally the same as in tension (when it is pulled). Since strain is dimensionless, E has the units of stress (MN/m^2 or GN/m^2). Typical values are given in Table 1 below.

TABLE 1. *Values of α and E at $0°C$*

Material	α	E
Aluminium	$22 \times 10^{-6}/°C$	$70 GN/m^2$
Beryllium	12	300
Brass	18	100
Molybdenum	5	320
Steel	11	210
Titanium	8	110
Zirconium	6	90

1.6. Statical Determinacy

In the preceding three sections we have discussed the three fundamental concepts—equilibrium, compatibility and environment-deformation relationships—which enable us to undertake the analysis of a structure. We must now turn to an important distinction which occurs between two main types of structural system.

Let us consider the process of constructing the braced framework shown in Fig. 8(a). The joints are loosely pinned, so that there is no restraint at a joint against relative rotation of the bars which are connected there.

We begin by assembling bars AB, BC, CD and DA (the reader can do this for himself with meccano, or a similar constructional toy). This assembly of four bars is termed a "mechanism". It has no particular shape—various possible configurations are shown in Fig. 8(b)—and if a load is applied to the mechanism, it collapses.

Now insert bar AC. If bar AC has its correct length of 25cm we obtain the frame shown in the first of the diagrams of Fig. 8(c). The frames obtained with members AC of lengths 20cm and 30cm are also shown. Whatever the length of AC (within wide limits) it can be inserted without difficulty into the mechanism ABCD. The frame

BASIC CONCEPTS

we have obtained from the mechanism of Fig. 8(b) by inserting the bar AC has a definite shape and it is capable of resisting load: it is termed a "statically determinate structure".

(a)

(b)

MECHANISM

(c)

STATICALLY DETERMINATE STRUCTURE

(d)

STATICALLY INDETERMINATE STRUCTURE

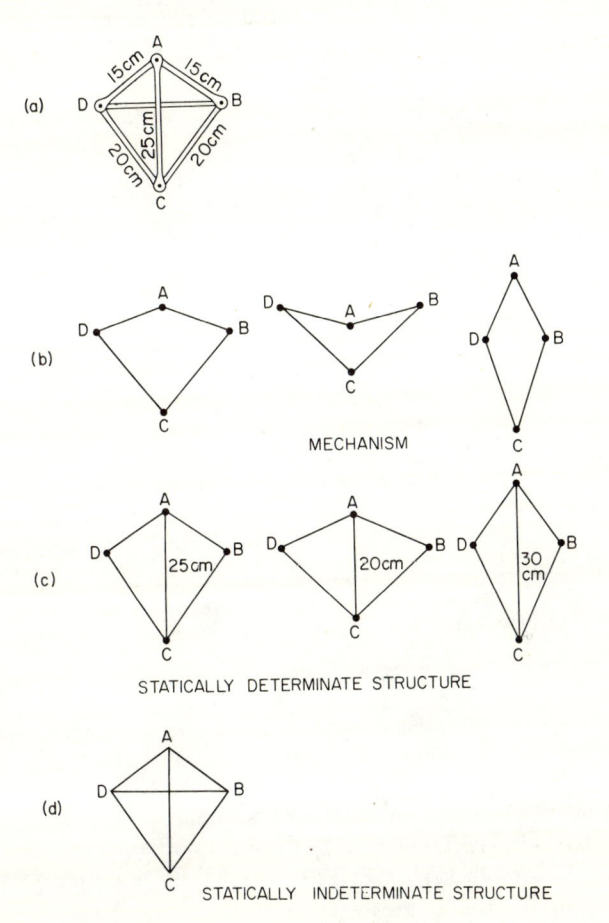

FIG. 8. Construction of framework

Now let us consider the insertion of the final bar DB into the frame shown in the first of the diagrams of Fig. 8(c). The length of the gap DB may be found by a simple calculation to be 24cm. If bar DB is exactly 24cm long it will fit into place without difficulty, but if its length differs from 24cm the bar will have to be forced into

15

BRACED FRAMEWORKS

position and stresses will be set up both in the bar itself and in the rest of the frame. A frame of this kind (Fig. 8(d)), which contains more bars than are necessary to make it statically determinate, and in which lack of fit of a bar would cause internal stresses to occur, is said to be a "statically indeterminate structure", or a "redundant structure".

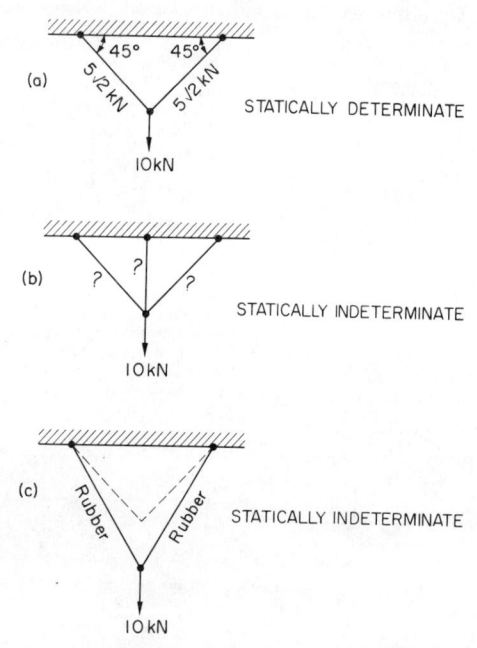

FIG. 9. Statical determinacy and indeterminacy

Although one can always determine whether a frame is statically determinate or indeterminate by the "building-up" process described above, the method itself gives little clue to the meaning of these phrases or the purpose of the distinction. This we shall now consider.

A statically determinate frame is one in which the forces in the bars due to a given set of applied loads can be found with sufficient accuracy from considerations of equilibrium alone. In the absence of applied loads the bar forces are zero. Figure 8(c) shows frames of this type. In Fig. 9(a) is another statically determinate frame consisting of two bars: consideration of vertical equilibrium shows that in each of them there is a tensile force of $5\sqrt{2}$kN. If we add

BASIC CONCEPTS

a third bar to the system (Fig. 9(b)) it is no longer statically determinate. We know that the sum of the vertical components of the forces in all three bars is equal to 10kN, but we cannot determine by consideration of equilibrium how much of the load is carried by each bar: this depends on the cross-sectional areas of the bars, on the stress–strain curves of their materials and on the initial lack of fit. To find the bar forces we must consider compatibility and material properties as well as equilibrium. A frame in which the bar forces cannot be found by consideration of equilibrium (or statics) alone is said to be statically indeterminate or redundant. The additional restraints over and above those necessary for the structure to be statically determinate (such as the force in the vertical bar in Fig. 9(b)) are known as "redundancies".

It is important to realise that the separation of equilibrium and compatibility, as is done in the analysis of a statically determinate structure, is essentially a process of approximation. No structure is truly statically determinate, but many structures are sufficiently so for practical purposes. If the bars shown in Fig. 9(a) were made of steel of 1cm^2 cross-section, the original angle of 45° would change to 45·02° when the load was applied. This change of geometry is clearly insignificant and the original angle may be used to determine the forces in the bars. If the bars shown in Fig. 9(a) were made of rubber, however, the change in geometry on application of the load might be large (Fig. 9(c)) and the structure could no longer be regarded as being statically determinate.

The phrase "statically determinate" is not meant to exclude dynamic problems. These may be reduced to the equivalent static system by the use of d'Alembert's principle and the considerations of this section may then be applied.

1.7. Methods of Structural Analysis

All methods of structural analysis depend only on the three fundamental considerations of equilibrium, compatibility and material properties, but the ways in which these concepts may most conveniently be used in the analysis vary from structure to structure. In particular, the analysis of a statically indeterminate structure, where the equilibrium and compatibility conditions cannot be

BRACED FRAMEWORKS

separated, differs from that of a statically determinate system, where equilibrium and compatibility can be treated separately.

In the present section we consider the various methods of structural analysis. The student will not find all that is written completely intelligible at a first reading, but he is urged to re-read this section at intervals as he becomes more familiar with later parts of the book.

The purpose of all structural analysis, as was explained in §1.2, is to determine whether a given structure can satisfactorily fulfil a given function. The first stage in the analysis of a structure must therefore be to define the function it is to fulfil. In examination questions the student is given the "loading conditions", but outside the examination hall these may be very difficult to determine. The load on a bridge due to its own weight (dead load) can be determined quite easily. The load due to present traffic (live load) can probably be estimated fairly accurately. The load due to the traffic twenty-five years hence is less certain, and the loads due to wind, ice formation and earth tremors may be less certain still. The loads on the floors of a block of flats may be reasonably estimated by weighings of typical furniture, but no one can be quite sure that all the tenants are not going to be bibliophiles who will fill their spare bedrooms with books to a depth of several feet. The estimation of the likely function of a structure is a most interesting problem of structural design, but we shall not discuss it further here. Henceforth we shall assume that the function has been specified and that we need only consider the resulting behaviour of the structural system.

In the case of a statically determinate system, no internal forces exist in the absence of applied loads. If loads are applied to the structure, we can proceed straight away, by considering equilibrium, to determine the internal forces. The methods which are used to find the forces in the bars of a framework are described in §2.4. Once the forces in the bars are known, it is possible from our knowledge of the material properties to determine whether any individual bar will fail. If the individual bars are satisfactory, we determine their elongations (from the material properties) and proceed to build up the deformations of the framework as a whole by considering geometrical compatibility. The method used—that of displacement diagrams—is described in §2.8. If the displacements of the whole structure are acceptable, the structure will function satisfactorily.

BASIC CONCEPTS

The process of analysis described above for a statically determinate structure is quite straightforward. It uses the three basic concepts separately in the order: (i) equilibrium, (ii) material properties, (iii) compatibility. The analysis of a statically indeterminate structure is less simple.

The first stage in the analysis of a statically indeterminate framework is usually to determine the number of redundancies, to choose those restraints which are to be regarded as redundancies, and to give them suitable algebraic symbols. The redundant restraints may be the forces in certain bars or the reactions at certain supports. The process is described in §3.2.

Having defined the redundant restraints as $R_1, R_2, \ldots$, the forces in all the bars of the frame can now be found, from considerations of equilibrium, in terms of the applied loads and $R_1, R_2, \ldots$ By using the environment–deformation relations for the material of construction it is next possible to find the elongations of the bars as functions of $R_1, R_2, \ldots$ The remainder of the problem consists of choosing such values for the redundancies $R_1, R_2, \ldots$, that the frame fits together—i.e. that the compatibility conditions are satisfied. Having found $R_1, R_2, \ldots$, the deformations of individual bars and of the whole framework can be determined in exactly the same way as for a statically determinate structure, and the criteria of satisfactory function applied.

The difficulty in analysing a statically indeterminate structure lies in determining $R_1, R_2, \ldots$, in such a manner that the compatibility conditions are satisfied. Various ways of tackling the compatibility problem are available. The most straightforward is by the direct use of geometry. The lengths of the bars are calculated in terms of $R_1, R_2, \ldots$ The size of the gaps in the frame into which the bars have to fit are calculated, also in terms of $R_1, R_2, \ldots$ The values of R_1 and R_2 are chosen so that in each case the length of the member is equal to the length of the gap. This method (solution by displacement diagrams) is described in §3.3. It has the advantages of being quite fundamental, of being physically obvious, and of being applicable to any kind of environment–deformation relationship. The disadvantage of the method is that it becomes unwieldy unless the number of redundancies is very small.

As an alternative to the direct use of geometry, a group of solutions known as energy methods may be used. In energy methods, the

BRACED FRAMEWORKS

compatibility conditions are disguised as a set of requirements imposed on certain differential functions of the energies of the system. The energy methods are generally much easier to apply than the direct use of geometry and they can conveniently be used for a larger number of redundancies. The various forms (virtual work, Castigliano's theorems, complementary energy) are described in §§3.4 and 3.5. The principal advantage of the energy approach is that awkward geometrical calculations are replaced by energy equations which are much easier to formulate. There are two main disadvantages: since the physical process of fitting the structure together is no longer apparent, it is very much easier to make mistakes than when using geometry directly; and some of the most commonly used energy methods can only be applied to linear systems.

Both the direct use of geometry and the use of energy methods effectively lead to a set of simultaneous equations (in general non-linear) equal in number to the number of redundancies $R_1, R_2, \ldots$ In a complicated structure the number of equations may be very large. Provided a suitable computer is available the equations need not present too much difficulty, but in the absence of a computer the solution of the equations can be very laborious. There have therefore been developed a number of methods for the solution of redundant structures which avoid both the direct consideration of geometry and simultaneous equations. These types of solution are known as relaxation methods. In them, the structure is prevented from deformation under the applied loads by the insertion of a suitable number of restraints: for a framework these might be imagined as a set of jacks at each joint. Each restraint is then relaxed in turn and the re-distribution of the forces in the framework and on the other restraints is noted. The relaxation is carried out systematically so that the restraining forces are gradually reduced to negligible values. These methods are described in §3.6. Relaxation methods have the advantage of a return to a physical process of solution and they are therefore less prone to error than energy methods. The methods are essentially numerical, and they cannot very conveniently be used for algebraic solutions.

The methods of structural analysis which have been described above are shown schematically in Fig. 10. It is not necessary in every problem to complete the process indicated by this diagram. If the bar forces alone are required, the last two lines will be ignored.

BASIC CONCEPTS

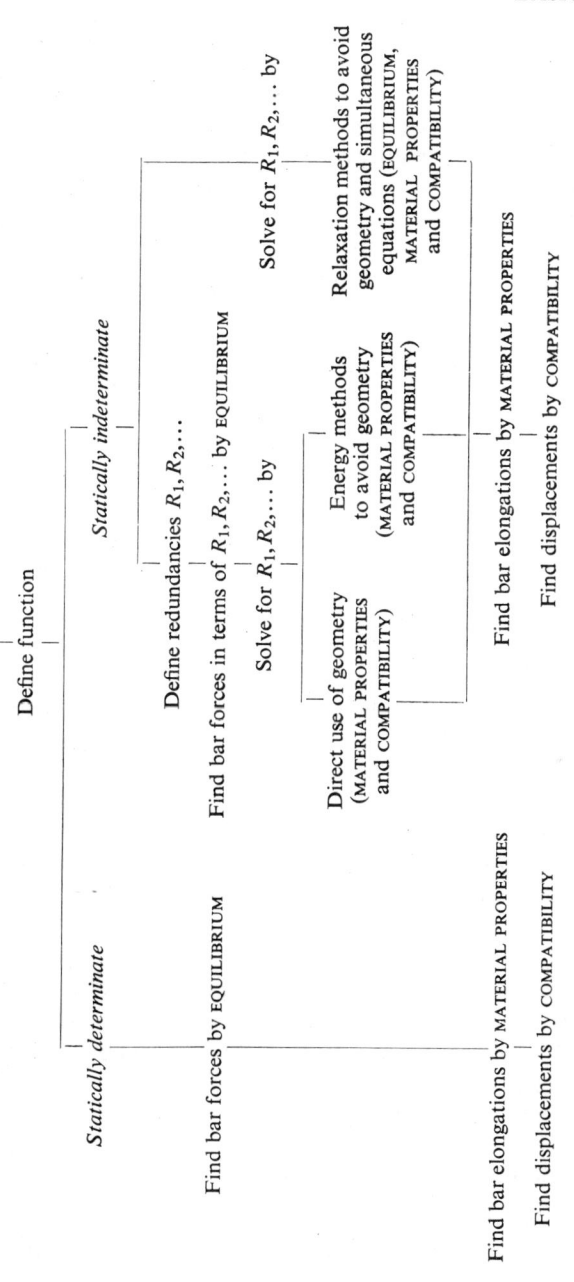

Analysis of a pin-jointed framework
Define function

Statically determinate
- Find bar forces by EQUILIBRIUM
- Find bar elongations by MATERIAL PROPERTIES
- Find displacements by COMPATIBILITY

Statically indeterminate
- Define redundancies $R_1, R_2, \ldots$
- Find bar forces in terms of $R_1, R_2, \ldots$ by EQUILIBRIUM
- Solve for $R_1, R_2, \ldots$ by
  - Direct use of geometry (MATERIAL PROPERTIES and COMPATIBILITY)
  - Energy methods to avoid geometry (MATERIAL PROPERTIES and COMPATIBILITY)
  - Relaxation methods to avoid geometry and simultaneous equations (EQUILIBRIUM, MATERIAL PROPERTIES and COMPATIBILITY)
- Solve for $R_1, R_2, \ldots$ by
- Find bar elongations by MATERIAL PROPERTIES
- Find displacements by COMPATIBILITY

FIG. 10. Analysis of a pin-jointed framework

1.8. Braced and Non-braced Frames

In this final section of Chapter 1 we define the class of structure with which this book is primarily concerned—the braced framework. It is important to realise, however, that the whole of the preceding chapter and much of what will follow in the further chapters is applicable with but slight modification to all kinds of structural system. Indeed, we would go further than that. Not only must the student avoid regarding beam problems as essentially different from those of braced frames, or problems on shafts as essentially different from those on beams; he must also avoid regarding structural analysis as a discrete and separate part of engineering science. The kind of thinking which is used in analysing the stresses in a bridge truss can also be used to find the currents in an electrical circuit or the temperature distribution in a steam turbine.

We proceed to distinguish between braced and non-braced frames. Consider the mechanism ABCD shown in Fig. 11(a). If a sideways load is applied to joint C, the mechanism collapses (Fig. 11(b)). We can prevent collapse—i.e. we can turn the mechanism into a structure —in one of two ways. If we replace one or more of the pin joints A, B, C and D by a rigid joint, the system will sustain load due to the resistance to bending of its members: examples are shown in Fig. 11(c). Alternatively, we can insert one or more additional bars (Fig. 11(d)): in this case the system sustains load because of the resistance to change of length of its members. A frame in which load is resisted by bending of the members is known as a non-braced frame, and one in which load is resisted by change of length of the members as a braced frame.

As may be inferred from Figs. 11(c) and (d), the deflections of a braced frame are small compared with the deflections of a comparable non-braced frame subjected to the same load. The student may demonstrate this for himself by comparing the stiffness of a thin strip of wood or card in bending, and under axial force—in addition, he may learn a great deal by making and testing cardboard models of the frames shown in Fig. 11. Because of this disparity in the order of magnitude of the deflections, making the joints of a braced frame rigid instead of pinned has negligible effect on the axial forces in the members: since the deflections are small, the bending of the members

BASIC CONCEPTS

associated with making the joints rigid is too small to make any significant contribution to balancing the applied load and this is almost entirely sustained by axial forces in the members, just as though the joints were still pinned. We shall return to this point in

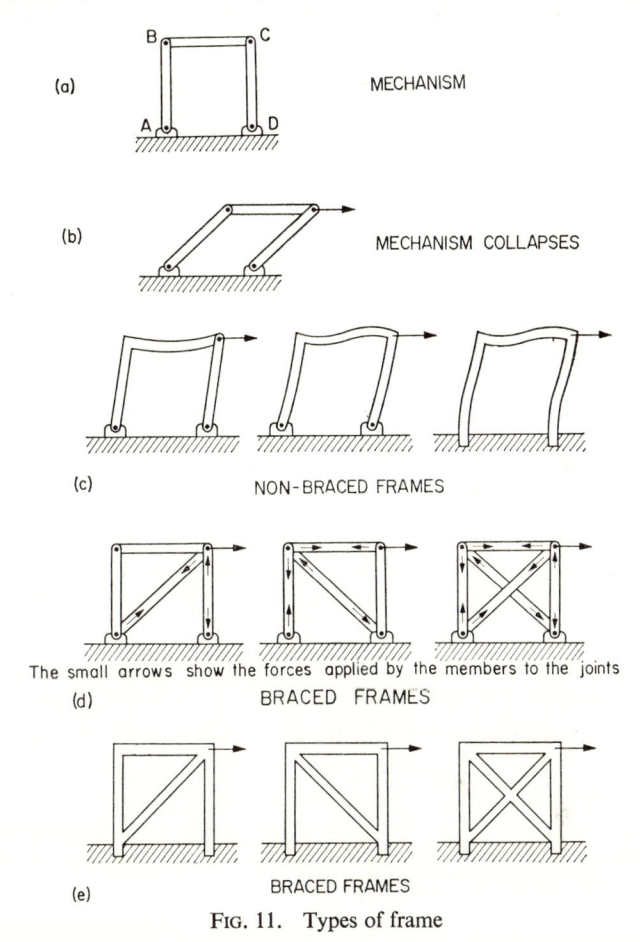

(a)   MECHANISM

(b)   MECHANISM COLLAPSES

(c)   NON-BRACED FRAMES

The small arrows show the forces applied by the members to the joints
(d)   BRACED FRAMES

(e)   BRACED FRAMES

FIG. 11. Types of frame

Chapter 4. All that we need note for the present is that the frames shown in Fig. 11(e), which have rigid joints, behave in essentially the same way as those in Fig. 11(d), which have pinned joints. The axial forces in the members and the deflections of the comparable frames in each set will be very closely the same.

BRACED FRAMEWORKS

The great majority of framed structures can be clearly divided into the braced and non-braced types, although structures in which part of the frame is braced and part is non-braced are not uncommon. Very occasionally one meets a framework which cannot be placed in

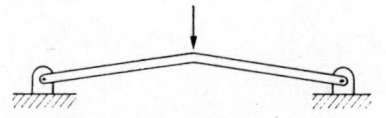

FIG. 12. Ill-conditioned frame

either of these categories—i.e. a structure in which both bending of a member and the axial force in it each make a significant contribution to supporting the applied load. Frames of this kind are usually associated with very ill-conditioned geometry in which the members at a joint are nearly parallel. An example is shown in Fig. 12. The analysis of ill-conditioned frames requires considerable care and skill.

EXAMPLES

1(a). Figure 13 shows the framework supporting the flat roof of a hangar. The doors, which are not shown, hang from the truss ABC. On erection, the clearance between the doors and the ground is 4·5cm. It is calculated that point B will

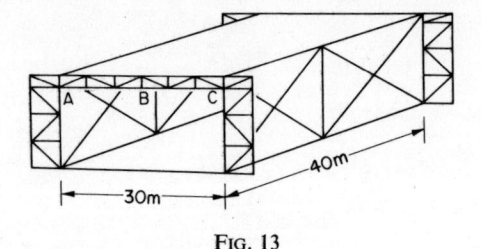

FIG. 13

descend by 0·5mm for each 1Mg of load uniformly distributed over the roof. The roof may be covered in winter by snow to a depth of 1m. Determine whether the structure will function satisfactorily. The density of snow may be assumed to be 160kg/m^3.

1(b). Sketch the forces acting on the following systems, making sure that they satisfy the requirement of equilibrium (the system to be considered is italicised in each case): an *aeroplane* in steady, level flight; a *railway engine* pulling a train at constant speed up an incline; the *string* of a kite; the *piston* of a petrol engine.

BASIC CONCEPTS

1(c). Determine which of the sets of measurements shown in Fig. 14 are geometrically compatible.

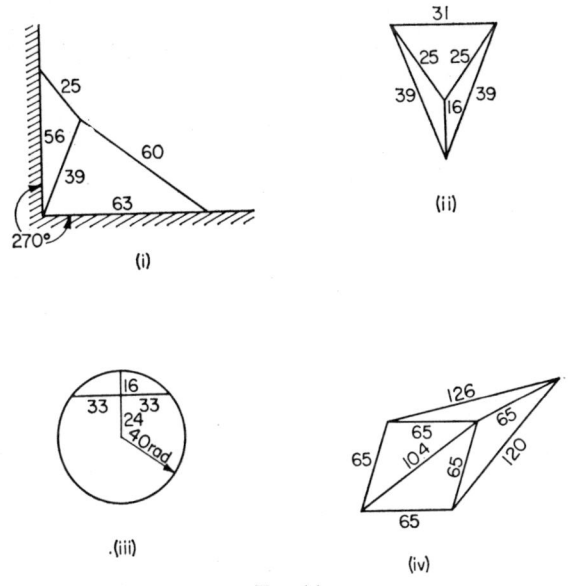

Fig. 14

1(d). A steel bar of 6cm×4cm rectangular cross-section is 3m long when free from load at 20°C. Find its length when subjected to a tensile force of 180kN at 80°C. Assume that the values of E and α given in Table 1 apply throughout the stress and temperature range.

1(e). An aluminium rod is fastened inside a titanium tube in such a way that the rod and tube must always be of equal length. The cross-sectional area of the rod is 13cm^2 and that of the tube, 23cm^2. The assembly is free from stress at 15°C. Determine the stress in the tube when both rod and tube are at 60°C. Assume that the values of E and α given in Table 1 are applicable.

1(f). The direct strain in a steel bar at a certain temperature can be written as

$$\varepsilon = 5 \times 10^{-6}\, \sigma + 10^{-8} \int_0^t \sigma \, dt$$

where σ is the stress in MN/m^2 and t is the time in hours. Find the strain when the bar is (i) subjected for 3000hr to a stress of 70MN/m^2; (ii) subjected for 3000hr to a stress of 40MN/m^2, which is then increased to 60MN/m^2; (iii) subjected to a stress which increases uniformly from zero to 60MN/m^2 over a period of 3000hr.

1(g). Find the true direct strains corresponding to direct strains of 0·00500, 0·0500 and 0·500.

25

BRACED FRAMEWORKS

1(h). The stress–strain curve of a certain material may be approximated by two straight lines: (1) of slope $7 \times 10^4 \text{MN/m}^2$ for stresses from 0 to 210MN/m^2; (2) of slope $2 \times 10^4 \text{MN/m}^2$ for stresses above 210MN/m^2. The unloading curve from any stress is a line parallel to (1). A bar of this material, originally of length 2m and cross-sectional area 20cm², is subjected to an axial force of 0·9MN which is subsequently reduced to 0·3MN. Find the new length of the bar.

1(i). Two identical bars AB and BC, each of length l, cross-sectional area A and Young's modulus E, are pinned together at B. The ends A and C of the bars are pinned to fixed points $2l$ apart, so that the bars lie in a straight line. Show that if a force P is applied at B in a direction perpendicular to AC, the lateral deflection of B will be $l(P/AE)^{1/3}$.

1(j). By sketching the construction of the pin-jointed frameworks shown in Fig. 15, determine which of them are statically determinate.

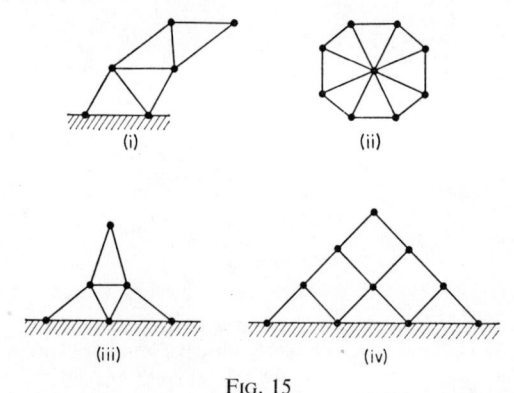

Fig. 15

1(k). The truss shown in Fig. 16 has rigid joints. It is subjected to a load as shown, and the central deflection can be measured. Describe the effects on the deflection of: (i) making all the joints into pin-points; (ii) removing the diagonal members; (iii) making all the joints into pin-points and removing the diagonal members.

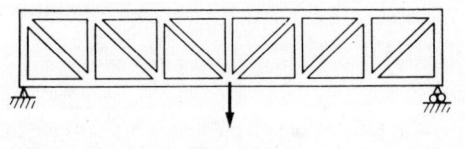

Fig. 16

2. statically determinate trusses

2.1. Introduction

A truss (or braced framework) is composed of members (or bars) connected together at joints (or nodes). The members of a truss are usually straight, but it is not essential to the function of the structure that they should be so. A truss in which the forces in the members due to external loading can be found by consideration of equilibrium alone is said to be statically determinate. It is characteristic of such structures that the bars can be assembled without forcing, even though they may not all be exactly of correct length. It follows that changes in length of the members of statically determinate trusses due to such causes as rise in temperature or humidity do not produce additional stresses in the structure.

2.2. Pin-joints and Pin-ended Members

The joints of a truss are usually rigid, the members being welded or riveted to a gusset plate (Fig. 17(a)), but as noted in §1.8 we shall in the present chapter regard the joints as pinned (Fig. 17(b)), whatever the mode of their construction. We do this because the behaviour

BRACED FRAMEWORKS

of a braced frame with rigid joints is essentially the same as that of one with pin joints, and the assumption that the joints are pinned greatly simplifies the analysis. The effects of rigid joints are discussed

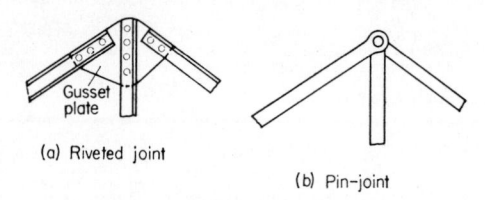

(a) Riveted joint

(b) Pin-joint

FIG. 17. Types of joint

in detail in Chapter 4. For three-dimensional frames we replace the idealised pin joints with spherical joints.

We proceed to investigate the equilibrium of a pin-ended member AB, connecting joints A and B (Fig. 18(a)). Since the connections are supposedly frictionless pins, no couple or moment can be trans-

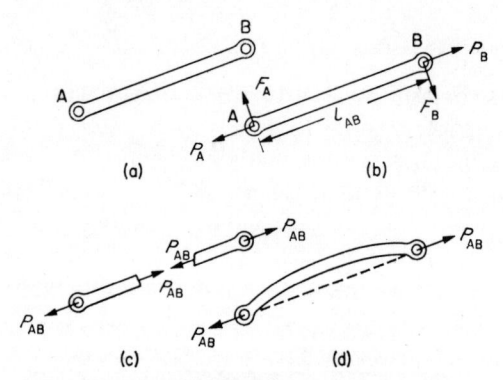

FIG. 18. Equilibrium of a pin-ended member

mitted to the member from either joint. Any force applied to the member at a joint can be resolved into components along and perpendicular to AB (Fig. 18(b)). Then if no external force is applied to member AB except through the pins, and we can ignore the weight of the member, we have, resolving along AB,

$$P_A = P_B \qquad (8)$$

Taking moments about A,

$$F_B \cdot l_{AB} = 0$$

whence
$$F_B = 0 \quad (9)$$

Resolving perpendicular to AB,

$$F_A = F_B = 0 \quad (10)$$

It follows from equations (8), (9) and (10) that the only possible force in a pin-ended member is a force applied along the line of the pins, and that at each end and at any cross-section the force has the same value, P_{AB} (Fig. 18(c)). The argument that we have been using does not depend upon the bar being straight—it can have any shape (Fig. 18(d)).

For a three-dimensional structure with bars connected by spherical joints, similar reasoning shows that the only possible force in a member is one applied along the line of the connections.

2.3. Equilibrium at Joints and Supports

Consider the joint shown in Fig. 19 in which all the members lie in one plane. Let there be n members, connected by a frictionless pin, making angles $\alpha_1, \alpha_2, \ldots, \alpha_n$ with the datum direction. Let the

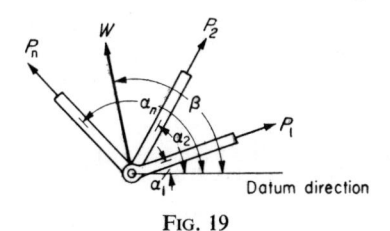

FIG. 19

forces in the members be $P_1 \ldots P_n$ and let a load W be applied to the joint at an angle β to the datum direction. Then under the action of all these forces the joint, which may be imagined to consist of the connecting pin, must be in equilibrium. Resolving in the datum direction, we have

$$P_1 \cos \alpha_1 + P_2 \cos \alpha_2 + \ldots P_n \cos \alpha_n + W \cos \beta = 0 \quad (11)$$

BRACED FRAMEWORKS

and resolving perpendicular to the datum direction,

$$P_1 \sin \alpha_1 + P_2 \sin \alpha_2 + \ldots P_n \sin \alpha_n + W \sin \beta = 0 \quad (12)$$

There are thus two independent equations of equilibrium between the forces at a joint in a plane truss. The equations can be obtained by resolving in any two directions—the directions do not have to be mutually perpendicular. It follows that if we know all of the bar forces except two at a given joint, these two bar forces can be found by considering the equilibrium of the joint. If the number of

(a) Fixed support   (b) Roller support

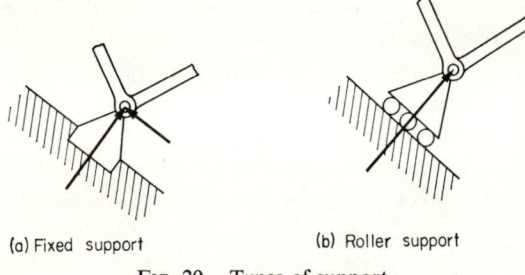

FIG. 20. Types of support

unknown forces at a joint exceeds two, that joint cannot for the present be solved.

Similar considerations apply when the joint is attached to a support of the structure. Supports can be of two kinds. The fixed support (Fig. 20(a)) is capable of resisting a force in any direction, and so for a plane truss the support can be imagined as applying two components of force to the joint. The roller support (Fig. 20(b)) can only apply one force to the joint in a direction perpendicular to that in which the rollers travel. If the number of unknown bar forces plus the number of unknown components of reactive force from the support does not exceed two, the joint can be solved.

For a three-dimensional frame, the number of independent equations of equilibrium at a joint is three and supports can be of three kinds: fixed (three components of force); single roller (two components of force); and double roller (one component of force).

STATICALLY DETERMINATE TRUSSES

2.4. The Forces in the Members of a Truss

2.4.1. *Successive resolution at the joints*. In the previous two sections we have considered the equilibrium of an individual bar and the equilibrium of a joint connecting a number of bars. We can now go on to consider the forces in the members of a complete structure. In §2.4 we shall tacitly assume that the structures with which we are concerned are statically determinate. The kind of solutions that we shall discuss, however, are equally useful when considering statically

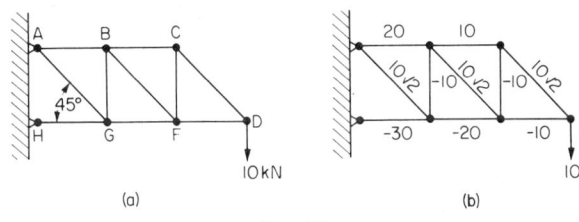

Fig. 21

indeterminate structures: the only difference is that in the latter case we shall be concerned with algebraic solutions rather than numerical ones. The conditions for statical determinacy are discussed in §2.5.

Before considering the methods of analysis which can be used to find the forces in the bars of a truss, we must first define a sign convention. If a bar is pulled, the force in it is said to be tensile and will be taken as positive. If a bar is pushed, the force in it is said to be compressive and will be taken as negative.

The simplest method of finding the forces in the bars of a plane truss is to extend the argument of §2.3 and to solve the structure by resolving the forces at each joint in turn. This method was devised by Whipple (1847), Jourawski (1850) and others. Consider the cantilever truss shown in Fig. 21(a). At joint D there are two unknown bar forces.

Resolving vertically at D, $10 - \sqrt{\tfrac{1}{2}} P_{CD} = 0$,
whence $P_{CD} = 10\sqrt{2}\,\text{kN}$.

Resolving horizontally at D, $P_{DF} + \sqrt{\tfrac{1}{2}} P_{CD} = 0$,
whence $P_{DF} = -10\,\text{kN}$.

31

BRACED FRAMEWORKS

We must next consider joint C, where there are now two unknown bar forces—those in BC and CF. We cannot yet proceed to joint F, since there are still three unknown bar forces there.

Resolving horizontally at C, $P_{BC} = 10\,\text{kN}$.

Resolving vertically at C, $P_{CF} = -10\,\text{kN}$.

We continue to analyse each joint in turn, choosing the route of our solution so that at each stage we are concerned with a joint having not more than two unknown bar forces. The route of

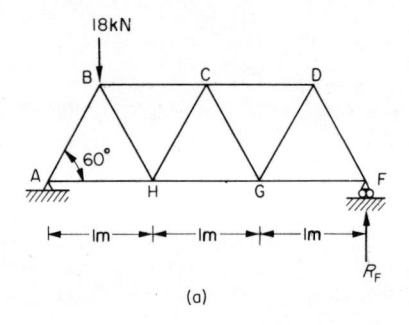

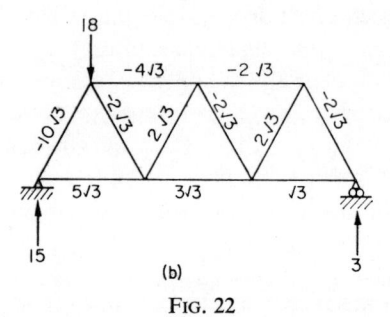

Fig. 22

solution is joint D → joint C → joint F → joint B → joint G. The forces in the bars, which the reader should check for himself, are shown in Fig. 21(b).

Sometimes it is necessary to consider overall equilibrium of the structure in order to find the reactions at the supports before analysing the joints. In the truss shown in Fig. 22(a) there are at least three unknown forces at each joint and it is not possible to "break in"

STATICALLY DETERMINATE TRUSSES

to the structure to begin the analysis. If we take moments about A, however, for the equilibrium of the whole truss, the reaction at F, which must be vertical since there is a roller support, can readily be found. We have

$$R_F \times 3 = 18 \times 0\cdot 5, \quad \text{whence } R_F = 3\text{kN}.$$

At joint F the number of unknown forces has now been reduced to two and we can complete the solution in the order joint F → D → G → C → H → B → A. The forces in the bars and the reactions at the supports, which the reader should check for himself, are shown in Fig. 22(b).

The method of solution by successive resolution at the joints is suitable for plane or three-dimensional trusses where the geometry is simple (e.g. 45° or 60° angles). If the geometry is complicated, however, the method can become very cumbersome and graphical solutions such as those described below may be preferable.

2.4.2. *Polygons of forces at the joints.* In the simplest graphical method, the process of successive resolution at the joints is replaced by drawing a polygon of forces for each joint in turn (Culmann, 1866). Consider the cantilever truss shown in Fig. 23(a). We begin at joint C, where there are two unknown bar forces, P_{BC} and P_{CD}. Figure 23(b) shows the triangle of forces for this joint. Since the arrows indicating the direction of the forces follow round the diagram in order, the force in bar BC is pulling the joint to the left—bar BC is therefore in tension. The force in bar CD is pushing the joint upwards and to the right—bar CD is therefore in compression. The magnitude of the forces can be measured from the diagram as $P_{BC} = +15\cdot 00\text{kN}$, $P_{CD} = -18\cdot 03\text{kN}$.

We proceed to joint D, where there are now two unknown bar forces, P_{BD} and P_{DF}. Bar CD is in compression: the force in it is therefore pushing joint D down and to the left. The triangle of forces for the joint is shown in Fig. 23(c). The force in bar DF is pushing joint D upwards and to the right—bar DF is therefore in compression. The force in bar BD is pulling joint D upwards and to the left—bar BD is therefore in tension. The magnitude of the forces can be measured from the diagram as $P_{DF} = -18\cdot 54\text{kN}$, $P_{BD} = 7\cdot 11\text{kN}$.

Finally, we consider joint B. The force polygon is shown in Fig. 23(d). The unknown bar forces are P_{AB} and P_{BF}. Their magnitudes can be measured from the diagram and their signs can be found by

33

BRACED FRAMEWORKS

considering the directions of the arrows. We obtain $P_{AB} = 20.62$kN, $P_{BF} = -2.27$kN.

The solution that we have given above is essentially the same as that of successive resolution at the joints, but with each joint solved graphically instead of by calculation. There is no unique polygon of

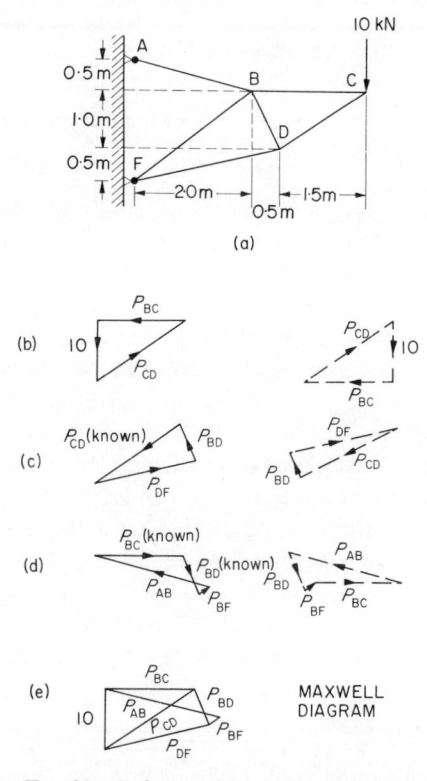

FIG. 23. Polygon of forces for each joint

forces at a joint—the diagram obtained depends upon the order in which the forces are taken. Alternative forms of force polygon are given in the broken-line diagrams of Fig. 23(b), (c) and (d).

2.4.3. *Maxwell diagrams and Bow's notation.* The method of solution given in §2.4.2 can be improved once it is realised that most of the forces occur in two diagrams. To avoid drawing each force twice, the diagrams can be superimposed. This is shown in Fig. 23(e).

STATICALLY DETERMINATE TRUSSES

A diagram such as Fig. 23(e) is known as a Maxwell diagram (Maxwell, 1864a) or on the Continent as a Cremona diagram (Cremona, 1872). A Maxwell diagram is quicker and easier to draw than the separate diagrams of Fig. 23(b), (c) and (d), but its use introduces two further problems. First, not all combinations of force polygons will superimpose successfully (those shown in full line in Fig. 23 will do so; those shown in broken line will not). Secondly,

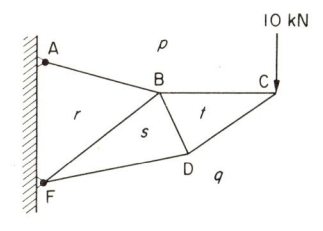

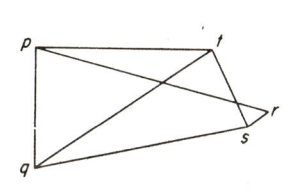

Member	Force	Value (kN)
AB	pr	+20·62
BC	pt	+15·00
CD	qt	−18·03
DF	qs	−18·54
BD	st	−7·11
BF	rs	−2·27

FIG. 24. Bow's notation

since the same bar force acts in opposite directions in the two polygons for the two joints at its ends, we cannot put directional arrows on the forces in a Maxwell diagram.

These problems can be overcome with the help of a device known as Bow's notation (Bow, 1873). Bow's notation consists of labelling each of the spaces bounded by bars of the structure or by external loads with a letter (minuscule letters are usually used, to avoid confusion with the joints of the truss, which are distinguished by capital letters). Each force is then identified by the two letters of the spaces which it separates. The Maxwell diagram is drawn by proceeding round each joint of the truss in turn, always in the same direction (say clockwise), and taking the forces in correct order. The

BRACED FRAMEWORKS

ends of the forces are labelled with the letters which identify them, and thus the junctions of the Maxwell diagram correspond to the spaces of the frame. If this scheme is followed, a consistent Maxwell diagram is necessarily produced. The signs of the forces can be

Member	Force	Value (kN)
AB	pv	−17·74
BC	rw	−16·00
CD	sw	+10·73
AD	tv	+18·34
BD	vw	+17·44

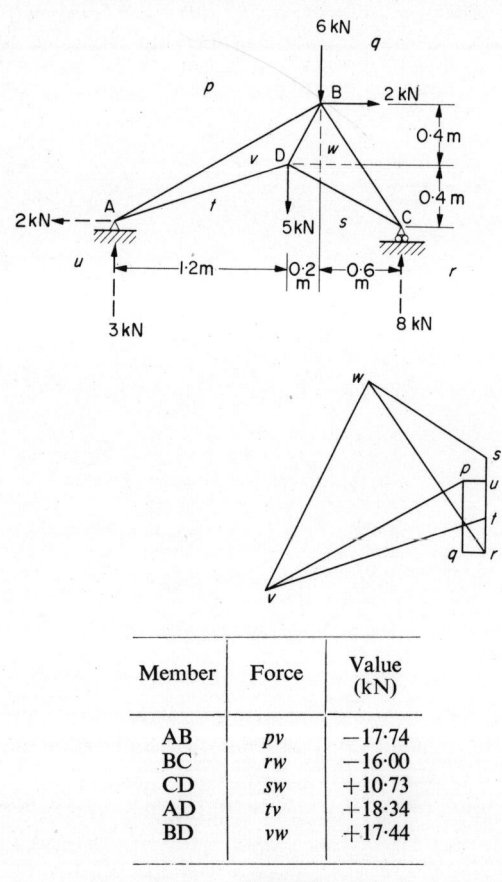

Fig. 25

found by considering the force polygon for each joint in turn. In any one polygon, the forces follow round in cyclic order—i.e. the (imaginary) directional arrows run in one consistent direction. The solution of the problem of Fig. 23 by Bow's notation is shown in Fig. 24.

STATICALLY DETERMINATE TRUSSES

Figure 25 shows a further problem, and its solution by the use of Bow's notation and a Maxwell diagram. The reactions at the supports have been found by considering overall equilibrium. As an example of the determination of the signs of the bar forces, consider joint B. The polygon of forces for this joint is *pqrwvp* (we proceed round the joint clockwise). The directional arrow to be assigned to *pq* is downwards, and the remaining arrows follow in cyclic order. Force *rw* is pushing joint B upwards and to the left: member BC is thus in compression. Force *wv* is pulling joint B downwards and to the left: member BD is thus in tension. Force *vp* is pushing joint B

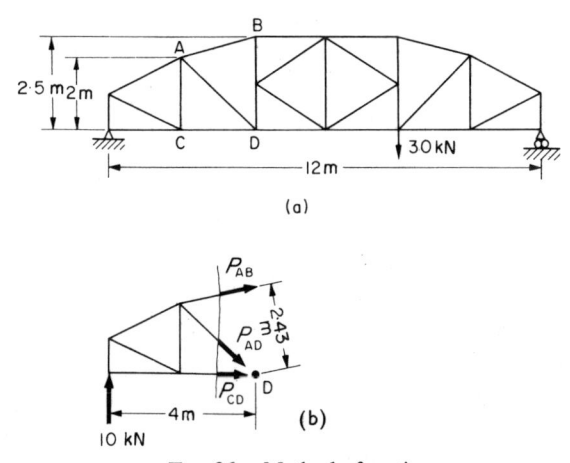

FIG. 26. Method of sections

upwards and to the right: member AB is thus in compression. The reader is strongly advised to re-draw the Maxwell diagram for this truss for himself, and to check the magnitudes and signs of the forces in all of the bars.

2.4.4. *Method of sections.* The methods of analysis described in §2.4.1–3 are very satisfactory when it is desired to find the forces in all the bars of the framework. If the forces in only a few of the bars are required, however, a more economical solution may be possible. Consider the plane truss shown in Fig. 26(a). Suppose we wish to find the force in member AB. The reactions at the supports can readily be found to be 10kN at the left-hand end and 20kN at the right-hand end. Now cut the truss through members AB, AD and

BRACED FRAMEWORKS

CD, and consider the left-hand portion (Fig. 26(b)). This part of the truss is in equilibrium under an external loading consisting of the 10kN applied at the support and the three unknown bar forces at the cut, P_{AB}, P_{AD} and P_{CD}. We can eliminate P_{AD} and P_{CD} by taking moments about the point at which they intersect, D. We have

$$10 \times 4 + P_{AB} \times 2 \cdot 43 = 0, \text{ whence } P_{AB} = -16 \cdot 5 \text{kN}.$$

P_{AD} can similarly be found by taking moments about the point where AB and CD intersect, or, now that P_{AB} is known, by resolving

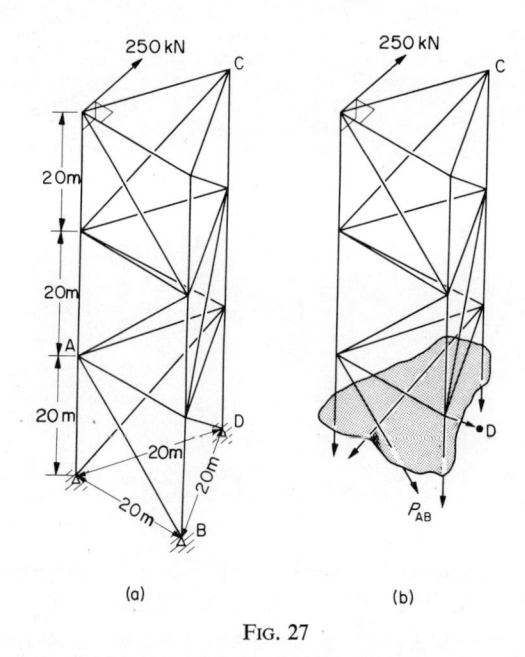

FIG. 27

vertically. The reader should try both methods for himself: the value of P_{AD} is $+8 \cdot 5$kN.

This type of solution is known as Ritter's method of sections (Ritter, 1862) although Schwedler (1851) also made a significant contribution to it. Since there are three independent equations of equilibrium for a plane body, the method of sections can be used provided the truss can be divided by a cut through the chosen bar and not more than two other bars: there are then three independent equations for the three unknown bar forces. The method of sections

STATICALLY DETERMINATE TRUSSES

could not be used to find the forces in any of the bars in the centre panels of the truss shown in Fig. 26(a) since a section would cut at least four bars. The bar forces in these panels can readily be found by any of the methods of §2.4.1–3.

In a three-dimensional system, where there are six independent equations of equilibrium for any part of the structure, the method of sections can be used provided a cutting surface can be found which passes through the chosen bar and not more than five other bars. As an example consider the triangular tower shown in Fig. 27(a) where it is desired to find the force in member AB. In Fig. 27(b) the structure has been divided by a surface passing through AB and five other bars. The force P_{AB} can be determined by considering moment equilibrium about the axis CD, since the remaining five bar forces either pass through CD or are parallel to it. We have

$$250 \times 10 = \frac{P_{AB}}{\sqrt{2}} \times \frac{20\sqrt{3}}{2} \quad \text{whence} \quad P_{AB} = +204\text{kN}.$$

2.4.5. Henneberg's method. The methods of determining the forces in the bars of a framework which have been described in §2.4.1–3 are of very wide application. Ritter's method of sections (§2.4.4) is limited to structures in which suitable cuts can be found, but for such structures it provides an excellent means of finding the forces in particular members. There are statically determinate frameworks, however, which cannot be solved by any of the methods of §2.4.1–4.

Consider the framework shown in Fig. 28(a). For simplicity, the interior bars may be imagined as crossing one another without connection. The forces in the bars of this structure cannot be found by any of our previous methods, since these methods depend upon finding a joint at which there are not more than two unknown forces. At each of the joints of the frame shown in Fig. 28(a) there are three unknown bar forces and so there is no joint at which we can "break in" to the frame to begin the analysis. The difficulty can be overcome by a variety of methods, notably those due to Henneberg (1886), Müller-Breslau (1887a), Saviotti (1888), Schur (1895), Mohr (1905) and Joukowski (1908). We shall describe that due to Henneberg.

Remove one bar of the frame (say AD) and replace it with another bar (say BG) such that the resulting structure can be analysed by one of our previous methods. Across the new gap AD apply two forces P_{AD} (Fig. 28(b)). Determine the forces in the bars in terms of the

BRACED FRAMEWORKS

external loads and P_{AD}. The frame can be solved by resolving at the joints in the order joint $D \to C \to F \to B \to A$. We obtain the values for the bar forces given in the second column of the table in Fig. 28(c). Now in fact there is no bar BG in the original frame, and so if

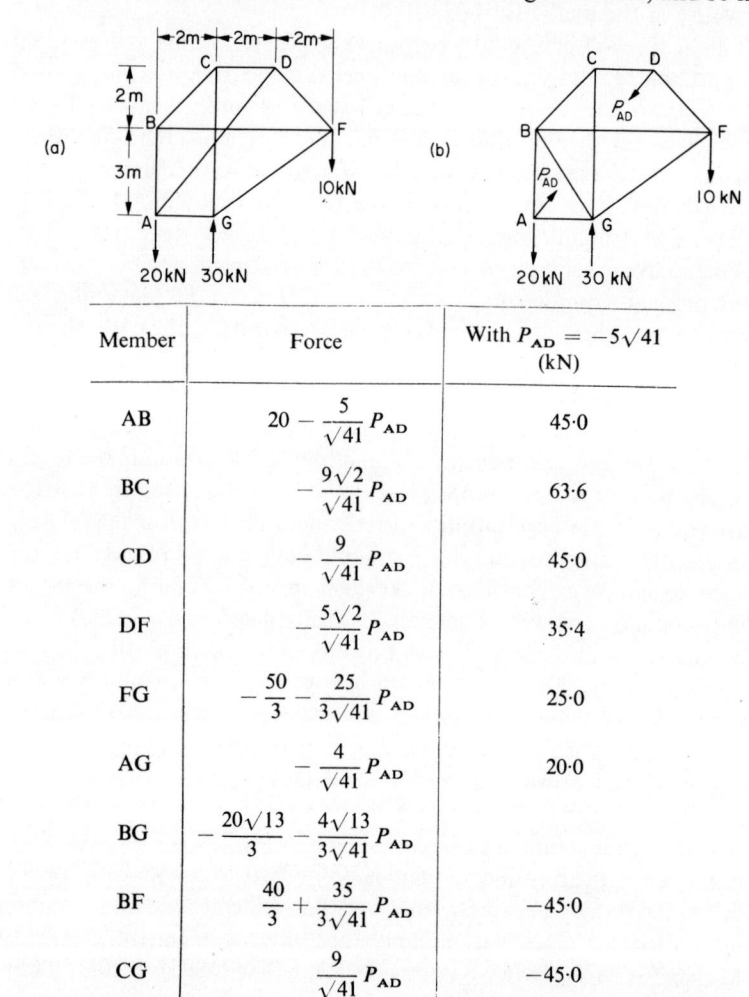

Member	Force	With $P_{AD} = -5\sqrt{41}$ (kN)
AB	$20 - \dfrac{5}{\sqrt{41}} P_{AD}$	45·0
BC	$-\dfrac{9\sqrt{2}}{\sqrt{41}} P_{AD}$	63·6
CD	$-\dfrac{9}{\sqrt{41}} P_{AD}$	45·0
DF	$-\dfrac{5\sqrt{2}}{\sqrt{41}} P_{AD}$	35·4
FG	$-\dfrac{50}{3} - \dfrac{25}{3\sqrt{41}} P_{AD}$	25·0
AG	$-\dfrac{4}{\sqrt{41}} P_{AD}$	20·0
BG	$-\dfrac{20\sqrt{13}}{3} - \dfrac{4\sqrt{13}}{3\sqrt{41}} P_{AD}$	
BF	$\dfrac{40}{3} + \dfrac{35}{3\sqrt{41}} P_{AD}$	−45·0
CG	$\dfrac{9}{\sqrt{41}} P_{AD}$	−45·0
AD	P_{AD}	−32·0

(c)

FIG. 28. Henneberg's method

the two systems of Fig. 28(a) and (b) are to be identical, the force in bar BG must be zero. We thus have

$$-\frac{20\sqrt{13}}{3} - \frac{4\sqrt{13}}{3\sqrt{41}} P_{AD} = 0$$

and so

$$P_{AD} = -5\sqrt{41} = -32 \cdot 0 \text{ kN}.$$

Substituting this value of P_{AD}, we can obtain the bar forces in the frame. These are given in the third column of the table of Fig. 28(c).

Henneberg's method is of use in solving certain lattice-type trusses (see Example 2(d)) but there are very few modern braced frameworks which cannot be solved by the analyses of §2.4.1–4, and the reader is unlikely to find Henneberg's method of much practical value.

2.4.6. *Tension coefficients.* The final method for the determination of the forces in the bars of braced frameworks which we shall

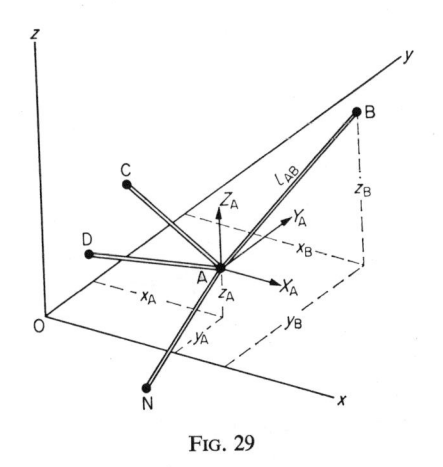

FIG. 29

describe is that of tension coefficients (Southwell, 1920). This method can be applied to any type of braced framework, but it is particularly useful when solving three-dimensional structures.

Consider joint A in a braced framework, connected by bars to a number of other joints B, C,...N (Fig. 29). Let the coordinates of A, B, C,..., N referred to Cartesian axes Ox, Oy, Oz be (x_A, y_A, z_A), (x_B, y_B, z_B), (x_C, y_C, z_C),..., (x_N, y_N, z_N). Then if the tensile force in bar

BRACED FRAMEWORKS

AB is P_{AB} and the length of bar AB is l_{AB}, we shall define a tension coefficient t_{AB} by the equation

$$t_{AB} = \frac{P_{AB}}{l_{AB}} \tag{13}$$

where $\quad l_{AB} = \{(x_B-x_A)^2 + (y_B-y_A)^2 + (z_B-z_A)^2\}^{\frac{1}{2}} \tag{14}$

The tension coefficients for each of the other bars are defined similarly as the force in the bar divided by its length.

Let the components of external load applied at joint A be X_A, Y_A, Z_A in the directions of the axes $0x$, $0y$, $0z$. Now the direction cosine of AB with respect to the coordinate axis $0x$ is $(x_B-x_A)/l_{AB}$. The component of the force P_{AB} in bar AB applied to joint A in the direction $0x$ is thus

$$P_{AB} \cdot \frac{(x_B-x_A)}{l_{AB}}$$

or, from equation (13), $\quad t_{AB}(x_B-x_A)$.

The components of the force P_{AB} applied to joint A in the $0y$ and $0z$ directions are similarly $t_{AB}(y_B-y_A)$ and $t_{AB}(z_B-z_A)$. The components of the force in bar AC applied to joint A in the directions $0x$, $0y$, $0z$ are $t_{AC}(x_C-x_A)$, $t_{AC}(y_C-y_A)$, $t_{AC}(z_C-z_A)$, and similarly for bars AD, ..., AN.

Resolving in the directions of the coordinate axes for the equilibrium of joint A we have the three equations

$$\left. \begin{array}{l} t_{AB}(x_B-x_A) + t_{AC}(x_C-x_A) + \ldots t_{AN}(x_N-x_A) + X_A = 0 \\ t_{AB}(y_B-y_A) + t_{AC}(y_C-y_A) + \ldots t_{AN}(y_N-y_A) + Y_A = 0 \\ t_{AB}(z_B-z_A) + t_{AC}(z_C-z_A) + \ldots t_{AN}(z_N-z_A) + Z_A = 0 \end{array} \right\} \tag{15}$$

Equations similar to (15) can be formulated for each joint of the frame. At supports, some or all of the X, Y, Z may be unknown. The equations can be written down very easily, since they involve only the projected lengths of the members on the coordinate axes, and this information can usually be taken directly from the drawing of the frame.

All of the equations do not have to be solved simultaneously—the tension coefficients can usually be found by solving small groups of equations in turn. Once the tension coefficients have been found, the

STATICALLY DETERMINATE TRUSSES

forces in the bars can be determined by using equations (13) and (14). We give an example below.

Consider the cantilever frame shown in Fig. 30, which is subjected to a load of 15MN at its tip. The positive directions of the coordinate axes are shown in the diagram, and the true lengths of the

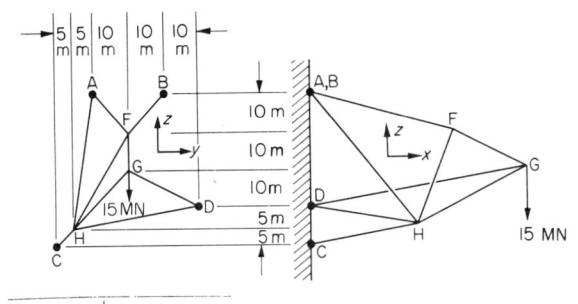

Member	True length (m)
AF	42·43
AH	46·37
BF	42·43
CH	30·82
DG	64·03
DH	46·37
FG	22·36
FH	30·82
GH	36·74

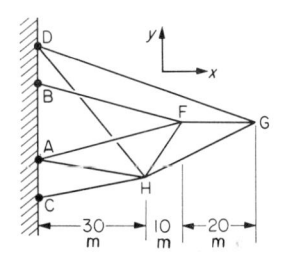

Fig. 30

members are tabulated. Resolving in the directions of the coordinate axes for the equilibrium of joint G, we obtain equations in a similar form to equations (15) as follows:

$$\left.\begin{array}{r}-20t_{FG}-60t_{DG}-30t_{GH}=0\\20t_{DG}-15t_{GH}=0\\10t_{FG}-10t_{DG}-15t_{GH}-15=0\end{array}\right\} \quad (16)$$

For joint F we have,

$$\left.\begin{array}{r}20t_{FG}-40t_{AF}-40t_{BF}-10t_{FH}=0\\-10t_{AF}+10t_{BF}-15t_{FH}=0\\-10t_{FG}+10t_{AF}+10t_{BF}-25t_{FH}=0\end{array}\right\} \quad (17)$$

BRACED FRAMEWORKS

and for joint H,

$$\begin{rcases} -30t_{AH} - 30t_{CH} - 30t_{DH} + 30t_{GH} + 10t_{FH} = 0 \\ 5t_{AH} - 5t_{CH} + 35t_{DH} + 15t_{GH} + 15t_{FH} = 0 \\ 35t_{AH} - 5t_{CH} + 5t_{DH} + 15t_{GH} + 25t_{FH} = 0 \end{rcases} \quad (18)$$

Equations (16), (17) and (18) can be solved in turn to yield the following values for the tension coefficients: $t_{DG} = -\frac{3}{16}$, $t_{FG} = \frac{15}{16}$, $t_{GH} = -\frac{1}{4}$; $t_{AF} = \frac{135}{352}$, $t_{BF} = \frac{45}{352}$, $t_{FH} = -\frac{15}{88}$; $t_{AH} = -\frac{117}{220}$, $t_{DH} = \frac{37}{440}$. Multiplying these tension coefficients by the true lengths of the members we obtain the bar forces in MN as follows: $P_{AF} = +16\cdot3$, $P_{AH} = +6\cdot5$, $P_{BF} = +5\cdot4$, $P_{CH} = -16\cdot4$, $P_{DG} = -12\cdot0$, $P_{DH} = +3\cdot9$, $P_{FG} = +21\cdot0$, $P_{FH} = -5\cdot3$, $P_{GH} = -9\cdot2$.

In the example above we have used the same system of coordinate axes for all joints. There is no necessity to do this, and different coordinate axes may be used at different joints if this makes formulation of the equilibrium equations easier. The use of tension coefficients enables three-dimensional frames to be solved in a straightforward manner: they can also be used for plane frames, but here they have no particular advantage over the other methods of solution which have been described previously.

Some interesting examples of three-dimensional frameworks will be found in Asplund (1966).

2.5. Conditions for Statical Determinacy

In all our work in Chapter 2 so far, we have assumed that the structures with which we were concerned were statically determinate. We must now investigate formally the conditions that this shall be so. As we have previously noted in §1.6, no structure is truly statically determinate, since this would imply no deformation under load, but many structures can be regarded as being statically determinate with sufficient accuracy for engineering purposes.

The first essential condition is geometrical: the shape of the framework must not change significantly throughout the range of its environment. Provided the changes of shape are small, we can still use the original angles when considering equilibrium at the joints: if the changes in the angles are significant, the bar forces are dependent on the deformations and the framework is no longer statically determinate. Knowing whether the geometrical changes in a given framework are likely to be significant is largely a matter of experience, although once the bar forces have been determined it is always possible to check one's assumptions. For frames composed of well-conditioned triangles (angles 20°–140°) made of the usual structural materials (in which the direct strains under load do not exceed 0·5 per cent) the changes in geometry of the structure are not likely to cause significant errors in values of the bar forces determined by using the original geometry. If the structure is ill conditioned, or the material unusually extensible, it may be necessary to consider the effects of the changes in geometry more carefully (see Fig. 9(c) and Example 1(i)). An interesting means of determining whether a framework is ill conditioned is given by Mobius (1837). Remove one member (say AB) from a statically determinate framework. Then the resultant assembly is a mechanism. If the configuration of the mechanism is such that distance AB is a maximum or minimum, then for small movements of the mechanism distance AB will not change. Member AB is thus incapable of preventing such movements, and the framework is ill conditioned.

The second condition for statical determinacy is that the forces in the bars of the structure must be independent of the material of construction and of the cross-sectional areas of the bars, and they must be unchanged by small variations in the lengths of the bars. As we saw in §1.6, this condition implies that the framework can be fitted together without forcing even though the bars are not quite of their correct lengths. It also implies that there are no initial stresses in the framework and that variations in length of the bars due to changes in temperature or humidity do not affect the forces in the bars. We can, in fact, determine whether a framework satisfies this condition for statical determinacy by sketching the process of construction (see §1.6 and Example 1(j)), but the condition can also be expressed algebraically (Mobius, 1837, and Maxwell, 1864b).

We begin by considering a plane frame composed of b bars. The frame will rest on a number of supports. These may be fixed supports providing two reactive forces, or roller supports providing one reactive force (see Fig. 20). Let the total number of reactive forces at the supports be r. Then since there are b unknown bar forces in the frame, the total number of forces to be determined for a complete solution is $b+r$. Suppose that the total number of joints in the frame (including those at the supports) is j. At each joint we have two independent equations of equilibrium, so that in total there are $2j$

BRACED FRAMEWORKS

independent equations relating the unknown bar forces and reactive forces. For statically determinacy, the number of available equations of equilibrium must be equal to the number of unknown forces. We thus have

$$b + r = 2j \qquad (19)$$

Equation (19) is a necessary condition for the statical determinacy of a plane frame: it is not a sufficient condition.

(a) $b = 10$
$r = 4$
$b+r = 14,\ 2j = 14$
STATICALLY DETERMINATE

(b) $b = 7$
$r = 5$
$b+r = 12,\ 2j = 12$
STATICALLY DETERMINATE

(c) $b = 8$
$r = 4$
$b+r = 12,\ 2j = 12$
STATICALLY DETERMINATE

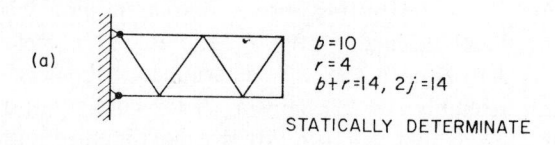

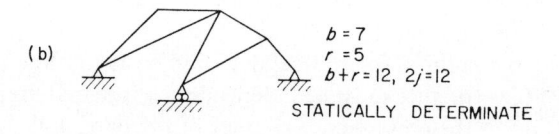

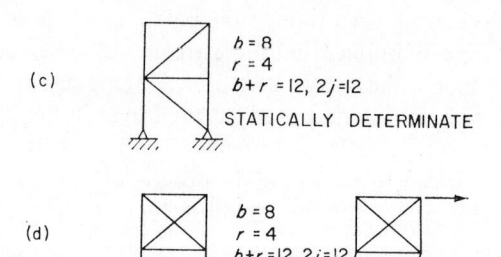

(d) $b = 8$
$r = 4$
$b+r = 12,\ 2j = 12$
MECHANISM

FIG. 31. Assemblies satisfying $b + r = 2j$

We illustrate the application of equation (19) with a number of examples. Consider the cantilever truss shown in Fig. 31(a). The truss has ten bars and two fixed supports, giving four reactive forces. $b+r$ is thus equal to 14. There are seven joints, so that $2j$ is also equal to 14 and the truss is statically determinate. It is important

STATICALLY DETERMINATE TRUSSES

to note that the lower support in Fig. 31(a) is a fixed support and is therefore regarded as providing two components of reactive force even though it is obvious by considering equilibrium at the joint that the force must act horizontally. In using equation (19) it is assumed that no prior consideration has been given to the equilibrium of the structure or its component parts, and in particular it is assumed that the reactive forces (two at each fixed support, one at each roller support) are unknown.

The frame shown in Fig. 31(b) has seven bars, five reactive forces and six joints: it is statically determinate.

The frame shown in Fig. 31(c) has eight bars, four reactive forces and six joints: it is statically determinate. The frame shown in Fig. 31(d) also has eight bars, four reactive forces and six joints, but it is not statically determinate. It is a mechanism which will collapse under side load, as shown in the right-hand diagram of Fig. 31(d). The upper panel of the frame is over-braced (statically indeterminate), whereas the lower panel is not braced and has become a mechanism. We must clearly add a further condition to equation (19)—for statical determinacy equation (19) must be satisfied *and* the bars must be properly arranged. It is not possible to give a simple definition of "proper arrangement", nor is it necessary to try to do so: the reader will quickly learn with experience to recognise parts of a structure which have been over-braced or inadequately braced. In cases of doubt the difficulty can always be resolved by sketching the process of erection of the frame: indeed, the reader is urged to use this method of determining whether a structure is statically determinate in preference to the use of equation (19), since by considering the geometrical problem of fitting the bars together he will learn far more about the behaviour of frameworks than he will do by blindly applying a formula.

For three-dimensional frames, where there are three independent equations of equilibrium at each joint, equation (19) is replaced by

$$b+r = 3j \qquad (20)$$

where each support may now provide one, two or three reactive forces (see §2.3). As an example we may consider the frame of Fig. 30, where there are nine bars, twelve reactive forces and seven joints: equation (20) is satisfied and the structure is statically determinate.

BRACED FRAMEWORKS

We may summarise the formal conditions for statical determinacy as follows:

(1) The shape of the framework must not change significantly throughout the range of its environment.
(2) The number of bars (b), reactive forces (r) and joints (j) must be related by the equation
$$b+r = 2j \quad \text{(plane frame)}$$
$$3j \quad \text{(three-dimensional frame)}$$
(3) The bars must be properly arranged.

It is important to notice that these conditions make no mention of the applied load system. Whether a framework is statically determinate depends only upon the geometry of the framework and its supports: provided condition (1) is satisfied, statical determinacy is in no way dependent on the loads applied to the structure.

2.6. Superposition of Loads on a Truss

It is the function of most structures to have to withstand more than one type of load distribution and often it is not immediately apparent which of the loading conditions is likely to provide the most severe criterion. Under these circumstances it is necessary to analyse the structure for each of a number of different loading conditions, and if every one of these analyses had to be started *ab initio*, the work could become very tedious. The labour of solution can often be reduced if it is possible to build up the loading cases in which one is interested from simpler loading systems. Consider the truss shown in Fig. 32(a), subjected to a load of 16kN. Suppose that we find the force in member AB due to this loading to be 5kN. Now consider the loading of 24kN shown in Fig. 32(b). Suppose that the force in AB due to this is 15kN. Finally, consider the loading shown in Fig. 32(c), which is the sum of those in Figs. 32(a) and (b). Under some circumstances it may be possible to say that the force in AB due to the loading system (c) is equal to the sum of the forces due to systems (a) and (b), i.e. $5+15 = 20$kN.

This adding together of the bar forces due to a number of separate loading systems is an example of a more general idea known as the principle of superposition. In our present context the principle of

STATICALLY DETERMINATE TRUSSES

superposition states that if we calculate the force in a particular bar of a framework due to a loading system a applied alone as P_a, and due to a loading system b applied alone as P_b, then if we apply the loading systems a and b simultaneously, the force in the bar will be $P_a + P_b$. It is important to realise that the principle of superposition is not of general validity—there are many structural systems for which it is untrue—but in those cases where it can be applied, the use of the principle of superposition often leads to considerable economy of labour. The idea can be extended to any number of loading systems.

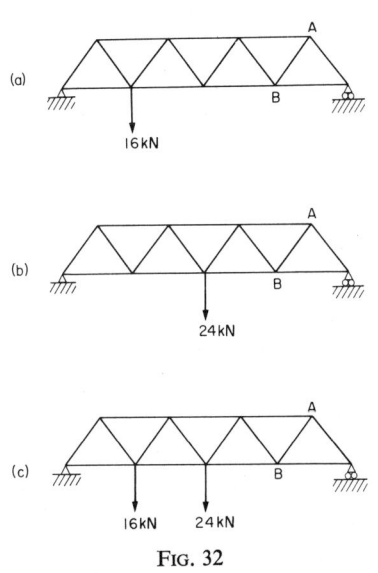

FIG. 32

If we consider a statically determinate truss subjected to a single load W_1, then since the geometry of the structure is not significantly changed by the application of load (by definition), all of the reactions and bar forces will be proportional to W_1. An alternative way of regarding this is to note that the size of the Maxwell diagram for the truss is proportional to W_1. In particular the force in bar AB may be written as $\alpha_1 W_1$, where α_1 is a constant determined only by the geometry of the truss and the position and direction of application of W_1. If we now consider the same truss under a load W_2, applied at some other point, the force in bar AB may be written as $\alpha_2 W_2$, where α_2 is again a constant. In general, if we have a number of

BRACED FRAMEWORKS

loads $W_1, W_2, W_3, \ldots$, applied separately, the bar force in AB will in each case be a linear function of the applied load. Provided that when we apply $W_1, W_2, W_3, \ldots$, simultaneously, the geometry of the structure remains significantly unaltered, the bar force in AB, which is derived from $W_1, W_2, W_3, \ldots$, only by geometrical relationships, will be equal to $\alpha_1 W_1 + \alpha_2 W_2 + \alpha_3 W_3 + \ldots$, and the principle of superposition will be valid. The applied loads and bar forces in a statically determinate truss form a linear system (see §1.5).

The only requirement for superposition of the bar forces due to a number of separate loading systems is that the geometry of the truss shall remain significantly the same whether the loads are applied individually or in combination. This, however, is one of the requirements that the truss shall be statically determinate (see §2.5). It follows that for a statically determinate truss the superposition of bar forces is always valid. For a redundant truss this is not generally true, and much more stringent conditions have to be observed if the principle of superposition is to be applied (see §3.1).

2.7. Influence Lines

One of the most important applications of the superposition of bar forces discussed in §2.6 is in the study of the effects of moving loads crossing a bridge. The analysis is greatly simplified by introducing the concept of an influence line, derived by Bresse (1865), Winkler (1868), Mohr (1868), Fränkel (1876) and Müller-Breslau (1887b). Consider the truss shown in Fig. 33 and suppose that we are interested in the force in member AB. The graph drawn underneath the truss

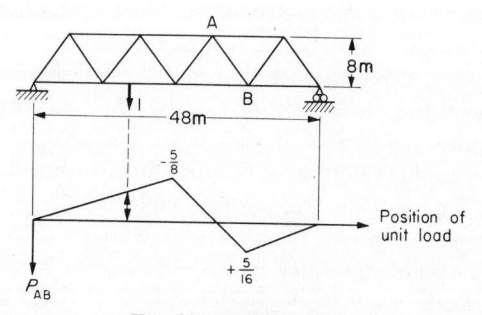

FIG. 33. Influence line

STATICALLY DETERMINATE TRUSSES

shows the variation of the force in AB with the position of a unit load on the bridge. A graph of this kind is known as an influence line. It is conventionally plotted with the bar force positive downwards.

So far in determining the bar forces in braced frameworks we have assumed that the loads were applied at the joints (or panel points, as

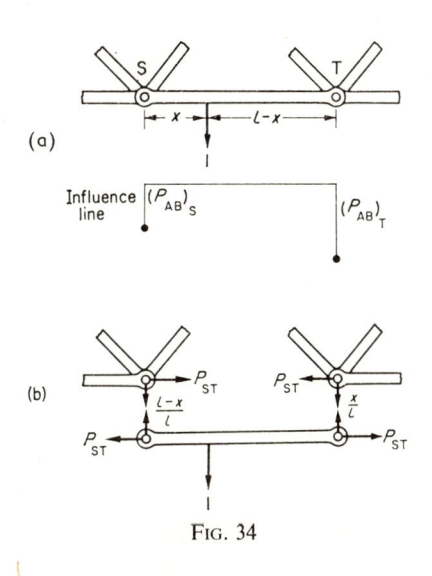

FIG. 34

they are also known). An influence line is, however, a continuous graph, and we must now consider the effect of applying load in between panel points. Suppose we have a unit load applied at distance x from end S of a bar ST of length l (Fig. 34(a)). Let the value of the force in bar AB when the unit load is at S be $(P_{AB})_S$ and when the unit load is at T let it be $(P_{AB})_T$. Now consider the equilibrium of bar ST (Fig. 34(b)). By taking moments about S and T the end reactions can be found to have the values $(l-x)/l$ and x/l, as shown. As far as the rest of the frame is concerned, applying unit load at x is equivalent to applying a load of $(l-x)/l$ at S and a load of x/l at T. By superposition, the force in bar AB is given by

$$(P_{AB})_x = \frac{l-x}{l}(P_{AB})_S + \frac{x}{l}(P_{AB})_T$$
$$= (P_{AB})_S + \frac{x}{l}\{(P_{AB})_T - (P_{AB})_S\}. \quad (21)$$

BRACED FRAMEWORKS

P_{AB} is a linear function of x as the unit load moves from S to T. The influence line is therefore a straight line between panel points. To draw the influence line we need only determine the value of P_{AB} when the unit load is at each panel point. The points on the graph so obtained can then be connected by straight lines. In fact it is not usually necessary to calculate P_{AB} for each panel point, but only for the points on either side of the panel containing the member concerned. The influence line usually falls linearly to zero at the ends of the truss from these two panel points (see Fig. 33). The determination of the force in member AB due to a unit load applied at a panel point can be done by any of the methods of §2.4 and we need

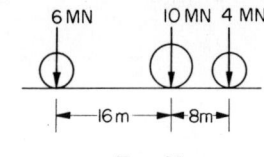

FIG. 35

not discuss the derivation of an influence line further; the reader is, however, advised to check the diagram of Fig. 33 for himself.

We illustrate the use of an influence line by considering the force in member AB of the truss shown in Fig. 33 when the truss is crossed by the train of loads shown in Fig. 35, which cannot be turned round. Since the influence line is linear between panel points, we need only consider values of P_{AB} when a load crosses a panel point: between these values the variation of P_{AB} with position of the train of loads will be linear. The complete calculation is shown in Fig. 36. The final graph in Fig. 36 shows the variation of P_{AB} with position of the 4MN load. It will be seen that member AB has to withstand a maximum tensile force of 1·88MN and a maximum compressive force of 7·50MN. If the train of loads could be turned round, we should have to repeat the calculation for the new arrangement and determine the absolute maximum values of the tensile and compressive forces in AB.

An influence line is also of great value in finding the force in a member of a truss due to a distributed load. Consider the bridge shown in Fig. 37 whose upper chord is to be crossed by a long uniformly distributed load of intensity w per unit length (the upper

STATICALLY DETERMINATE TRUSSES

members of a bridge truss are known as the upper chord and the lower members as the lower chord: the train of loads shown in Fig. 36 is running on the lower chord). Suppose that we are interested

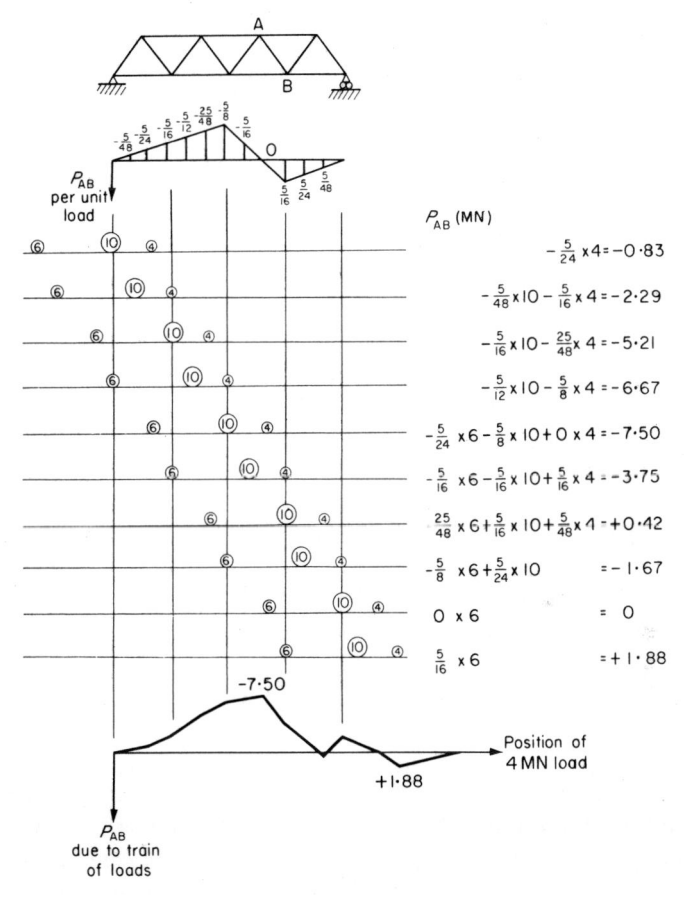

FIG. 36

in the force in member AB and that the influence line for this member is as shown. Let the load extend on to the bridge a distance d, and consider that part of the load between x and $x+\delta x$. This element of load has a value $w\delta x$: if the height of the influence line at x is h, the force in AB due to $w\delta x$ will be $hw\delta x$. The total force in AB

BRACED FRAMEWORKS

due to all of the distributed load between $x = 0$ and $x = d$ will be

$$P_{AB} = \int_0^d hw\,dx = w\int_0^d h\,dx$$

Now $\int_0^d h\,dx$ is the area under the influence line between $x = 0$ and $x = d$ (shown shaded in Fig. 37). It follows that we may write

$P_{AB} = w \times$ area under influence line for part of truss
covered by the load   (22)

As the load moves on to the bridge from the left, the area under the influence line increases until d is equal to 36m. Thereafter the influence line changes sign and the area decreases. If w were equal to 0·6MN/m, the maximum tensile force in AB would be

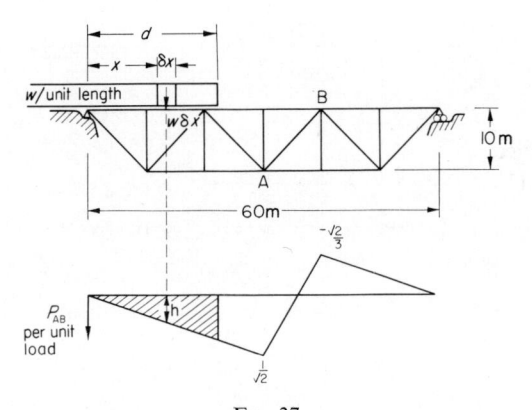

FIG. 37

$0.6 \times \frac{1}{2} \times 36 \times 1/\sqrt{2} = +7·6$MN (see Fig. 38(a)). To find the maximum compressive force in AB we must consider the load covering the whole of the negative part of the influence line (Fig. 38(b)). We find $P_{AB} = -0.6 \times \frac{1}{2} \times 24 \times \sqrt{2}/3 = -3·4$MN.

If the distributed load is not sufficiently long to cover the whole of the positive or negative areas of the influence line, it is necessary to find the position of the load for maximum force in the member. For a simple uniformly distributed load, the maximum occurs when the ordinates of the influence line at either end of the load are equal

STATICALLY DETERMINATE TRUSSES

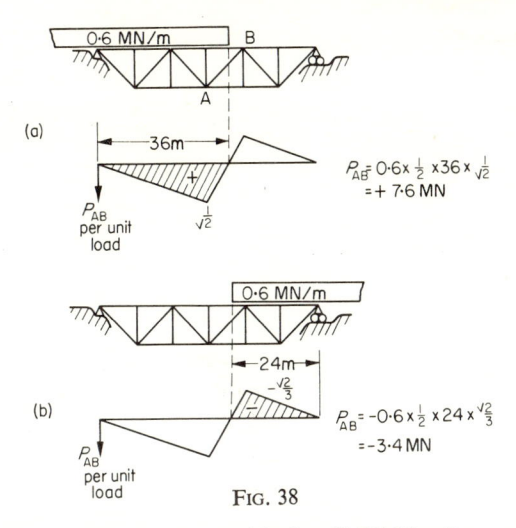

FIG. 38

(the reader may care to prove this for himself). For non-uniform distributed loads or trains of loads which are partly distributed and partly concentrated, the position of the loads to give the maximum bar force is usually most simply obtained by trial and error.

The subject of influence lines is discussed at some length by Matheson (1971).

2.8. The Displacements of Trusses

2.8.1. *Displacement diagrams.* We have now discussed at some length the determination of the forces in the bars of a statically determinate braced framework. Knowing the force in a bar, its temperature, humidity, etc., we can find the change in length of the bar from the environment–deformation relationships of the material from which it is made (§1.5). The next problem to be considered is how to determine the displacements of the joints of the framework from our knowledge of the change in length of the members. The problem is entirely geometrical. The ability to find the displacements of statically determinate trusses is important in itself (since these displacements may decide whether the structure functions satisfactorily) but it is also an essential step in the analysis of redundant trusses (see §3.3).

We begin by considering plane frameworks. Suppose we have two

55

BRACED FRAMEWORKS

bars AB and AC connected to each other at A and to two fixed points B and C, as shown in Fig. 39(a). Let the initial lengths of the bars be l_{AB} and l_{AC}. Now let the length of bar AB increase to $l_{AB} + \delta l_{AB}$ and that of bar AC to $l_{AC} + \delta l_{AC}$. We wish to find the new position of A. One way of doing this would be to plot the positions of B and C on drawing paper and to draw circular arcs of radii $l_{AB} + \delta l_{AB}$ and $l_{AC} + \delta l_{AC}$ about B and C. The intersection of these arcs would then give the new position of A, A' (Fig. 39(b)). Although this method would seem to provide a means of determining the displacement of A, in practice it is useless. The reason is that the changes in lengths of the members are usually very small (direct strains in steel structures are commonly less than 0·0005) and it is not possible to draw with sufficient accuracy to distinguish A' from A. The difficulty can be overcome by considering Figs. 39(a) and (b) superimposed. This has been done in Fig. 39(c), where the changes in length of the members have been exaggerated for clarity. Now because the changes in lengths of the members are in fact very small, the circular arcs in Fig. 39(c) approximate very closely to straight lines perpendicular to the directions BA and CA. We may therefore re-draw that part of the diagram consisting of the extensions of the members and the circular arcs as shown in Fig. 39(d). Figure 39(d) shows the displacement of joint A, and it is known as a displacement diagram, or Williot diagram (Williot, 1877). Since the diagram involves only the changes in length of the members it can be drawn to an exaggerated scale (100 times full size is commonly used) and the displacements can be determined accurately. It is usual to label the points on a displacement diagram with minuscule letters corresponding to the capital letters denoting the joints of the frame. Since the displacement diagram shows relative displacements, all fixed joints of the frame correspond to one point on the displacement diagram. (Figure 39(d) is re-drawn with the usual notation in Fig. 39(e): this shows the displacement of A relative to the fixed points B and C.)

We shall now construct the displacement diagram for the cantilever truss shown in Fig. 40(a). The extensions of the members are tabulated. We begin at the fixed point a, f. Bar AB extends by 0·062cm. B therefore moves away from A in the direction AB by 0·062cm (shown as ab' in the displacement diagram) and an unknown amount perpendicular to AB—i.e. b lies somewhere on a line through b'

STATICALLY DETERMINATE TRUSSES

perpendicular to ab'. Bar FB extends by 0·041cm. B therefore moves away from F in the direction FB by 0·041cm (shown as ab'' in the displacement diagram) and an unknown amount perpendicular to FB—i.e. b lies somewhere on a line through b'' perpendicular to fb''.

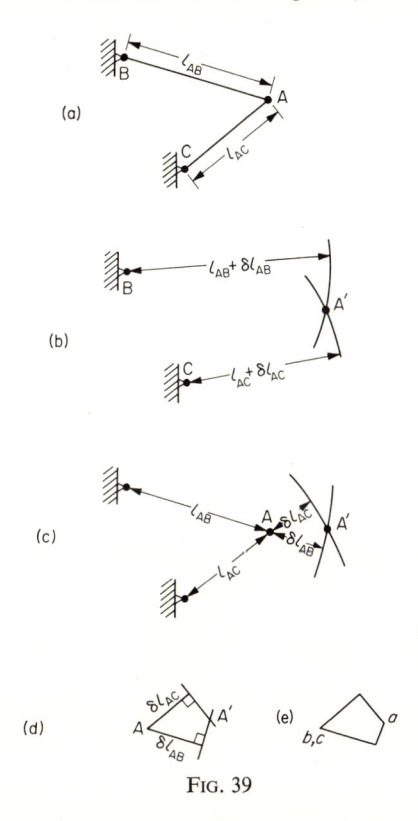

FIG. 39

The intersection of the line through b' perpendicular to ab' with the line through b'' perpendicular to fb'' gives the position of b.

Having found b, we can now determine the position of d. DF contracts by 0·056cm, so D moves 0·056cm relative to F in the direction DF (shown as fd') and an unknown amount perpendicular to DF. BD contracts by 0·055cm, so D moves 0·055cm relative to B in the direction DB (shown as bd'') and an unknown amount perpendicular to BD. The intersection of $d'd$ with $d''d$ gives the position of d.

57

BRACED FRAMEWORKS

The reader is left to deduce the position of C from those of b and d for himself. The completed diagram is shown in Fig. 40(b). The displacements of the joints can be measured from the diagram. That of C, for example (c relative to a, f), is 0·313cm down and 0·049cm to the left.

(a)

Member	Extension (cm)
AB	0·062
BC	0·037
CD	−0·033
DF	−0·056
FB	0·041
BD	−0·055

(b)

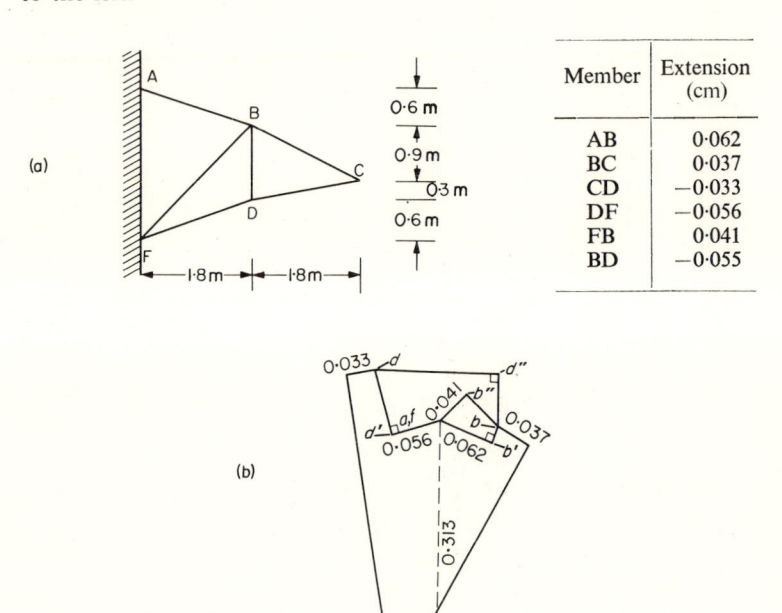

Fig. 40. Displacement diagram for cantilever

In the preceding example the cantilever truss was supported from two fixed points A and F each of which was connected to the same joint of the truss, B. We could therefore begin the displacement diagram at a, f, find the position of b and then proceed from joint to joint through the truss. If the truss is supported from two points which are not close together, however, it is more difficult to begin the displacement diagram. Consider the truss shown in Fig. 41(a), supported at A and F. A and F are not connected directly to a common joint and so there is no obvious means of beginning to

STATICALLY DETERMINATE TRUSSES

draw the Williot diagram. All that we can do is to make some initial assumption, draw the displacement diagram, and try to correct any errors which arise. The Williot diagram shown in Fig. 41(b) has been drawn by assuming that CH remains vertical (the letters in bold type should be ignored for the present). Starting from c the position of h

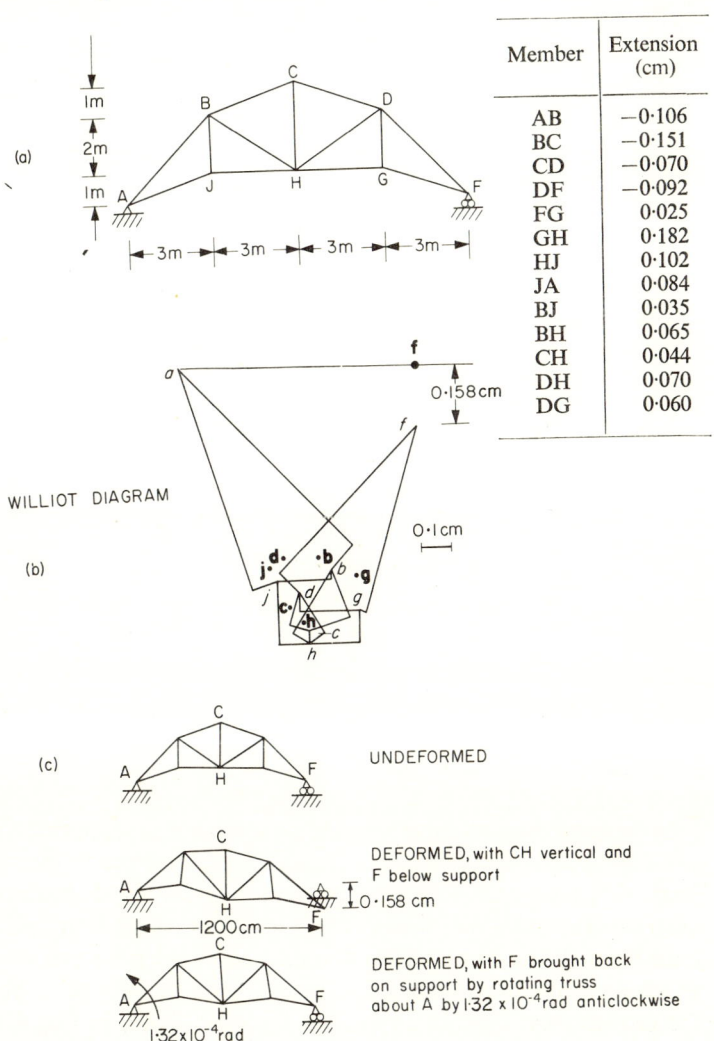

Member	Extension (cm)
AB	−0·106
BC	−0·151
CD	−0·070
DF	−0·092
FG	0·025
GH	0·182
HJ	0·102
JA	0·084
BJ	0·035
BH	0·065
CH	0·044
DH	0·070
DG	0·060

Fig. 41(a)–(c). Displacement diagram for simply supported truss

Joint	To the left (cm)	Upwards (cm)
B	0·040	0·040
C	0·053	0·079
D	0·040	0·119
J	0·013	0·040
H	0·013	0·079
G	0·013	0·119
F	0	0·158

FIG. 41(d). Displacements of joints due to rigid-body rotation about A of $1·32 \times 10^{-4}$ rad anticlockwise.

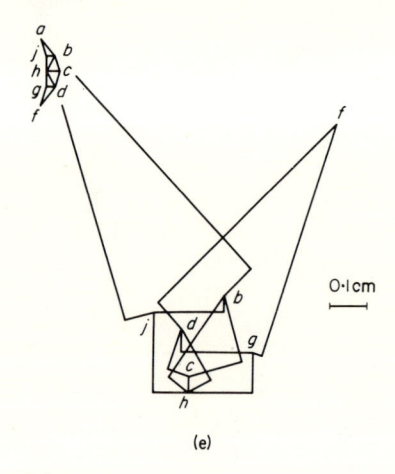

FIG. 41(e). Williot–Mohr diagram for simply supported truss

can then be plotted and the displacement diagram built up by proceeding via joints D, G, F and B, J, A.

By considering points a and f in the completed diagram it will be seen that our original guess that CH remains vertical was incorrect. A and F should be on the same level, but according to the Williot diagram of Fig. 40(b), F is 0·158cm below A. The deformations of the truss have been accounted for correctly, but the whole truss has suffered a clockwise rotation about A of $0·158/1200 = 1·32 \times 10^{-4}$ rad (see Fig. 41(c)). F can be brought back to its correct level by rotating the whole truss about A by $1·32 \times 10^{-4}$ rad anticlockwise.

STATICALLY DETERMINATE TRUSSES

The displacements of the joints due to this rotation (during which the truss is regarded as a rigid body) must be added to those of Fig. 41(b) in order to obtain the correct displacement diagram. The displacement of any joint due to the rigid-body rotation about A is equal to $1 \cdot 32 \times 10^{-4}$ rad multiplied by the distance between A and the joint, and it occurs in a direction perpendicular to that from A to the joint, but it is usually easier to calculate the displacement in vertical and horizontal coordinates. Due to an anticlockwise rotation of $1 \cdot 32 \times 10^{-4}$ rad about A, any joint will move to the left by $1 \cdot 32 \times 10^{-4}$ times its distance above A, and upwards by $1 \cdot 32 \times 10^{-4}$ times its distance to the right of A. The corrections calculated in this way are shown in the table of Fig. 41(d). By adding these displacements to those of Fig. 41(b) we obtain the final corrected diagram shown by the bold letters.

An alternative means of making the correction for rigid-body rotation about A is available. The additional displacements in Fig. 41(b) were due to an anticlockwise rotation about A of $1 \cdot 32 \times 10^{-4}$ rad. If, instead of adding the rigid-body displacements to the Williot diagram, we draw a new displacement diagram having the same fixed point a and showing the rigid-body displacements with all their signs reversed (i.e. the displacements due to a *clockwise* rotation of $1 \cdot 32 \times 10^{-4}$ rad), then the difference between corresponding points in the two displacement diagrams will give the correct absolute displacements. The diagram showing the rigid-body displacements with their signs reversed is known as a Mohr diagram: it is similar in shape to the framework but is perpendicular to it. The Williot and Mohr diagrams together are known as a Williot–Mohr diagram. The Williot–Mohr diagram for the truss under discussion is shown in Fig. 41(e).

Under some circumstances it may be possible to draw the correct displacement diagram *ab initio* even though the supporting points are not close together. In a symmetrical structure, symmetrically supported, whose members suffer symmetrical extensions, the centre member does not rotate and it can be used to provide the two starting-points of the diagram. We could not use this method in the problem above, since although the structure was symmetrical, the extensions were not. The properties of symmetrical systems will be discussed more fully in §2.10.

In some kinds of structure, the supports may deflect when the environmental conditions change. Provided these deflections are known, they can be incorporated in the displacement diagram without difficulty.

2.8.2. *Algebraic solution.* Displacement diagrams provide an excellent means of determining the displacements of the joints in plane frames, but for three-dimensional structures a direct algebraic solution is simplest if the displacements of all of the joints are required. Consider a member AB connecting joints A and B of a framework. Let the positions of the joints be defined by Cartesian coordinates and let $x_{AB} = x_B - x_A$, $y_{AB} = y_B - y_A$ and $z_{AB} = z_B - z_A$. The original length of member AB is thus $l_{AB} = (x_{AB}^2 + y_{AB}^2 + z_{AB}^2)^{\frac{1}{2}}$. Let the relative displacement of joint B from joint A in the x-direction be $u_{AB} = u_B - u_A$, in the y-direction $v_{AB} = v_B - v_A$ and in the z-direction $w_{AB} = w_B - w_A$, where u_A, etc., are the absolute displacements of the joints. Then the new length of member AB is

$$l_{AB} + \delta l_{AB} = \{(x_{AB} + u_{AB})^2 + (y_{AB} + v_{AB})^2 + (z_{AB} + w_{AB})^2\}^{\frac{1}{2}}$$

Remembering that the relative displacements are very small compared with the coordinate differences and that terms in u_{AB}^2, etc., may therefore be neglected, this equation may be rewritten as

$$l_{AB} + \delta l_{AB} = \{x_{AB}^2 + 2x_{AB} u_{AB} + y_{AB}^2 + 2y_{AB} v_{AB} + z_{AB}^2 + 2z_{AB} w_{AB}\}^{\frac{1}{2}}$$

$$= \{l_{AB}^2 + 2(x_{AB} u_{AB} + y_{AB} v_{AB} + z_{AB} w_{AB})\}^{\frac{1}{2}}$$

$$= l_{AB} \left\{1 + \frac{(x_{AB} u_{AB} + y_{AB} v_{AB} + z_{AB} w_{AB})}{l_{AB}^2}\right\}$$

We thus have

$$l_{AB} \cdot \delta l_{AB} = x_{AB} u_{AB} + y_{AB} v_{AB} + z_{AB} w_{AB}$$

$$= x_{AB}(u_B - u_A) + y_{AB}(v_B - v_A) + z_{AB}(w_B - w_A) \quad (23)$$

An equation similar to (23) can be written down for each bar of the frame. If we suppose that r of the joint displacements are defined (corresponding to the r reactive forces in the equilibrium problem) and that there are j joints, then $3j - r$ displacements have to be found for a complete solution. We have b independent equations, where b is the number of bars, so that the condition for solution to be possible is

$$b = 3j - r$$

STATICALLY DETERMINATE TRUSSES

This is same condition as that for statical determinacy (equation 20). If b exceeds $3j-r$, the structure is indeterminate and a consistent solution of equations (23) will only be possible provided the bar elongations form a compatible set.

We illustrate this method of solution by considering the framework shown in Fig. 42. The coordinate differences can be obtained from

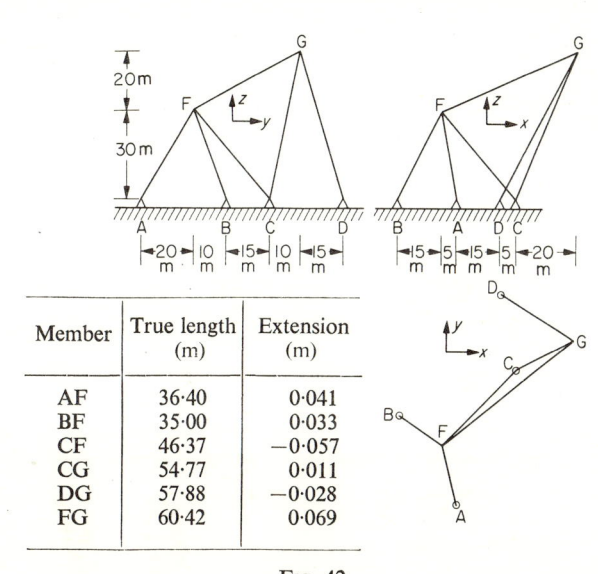

Member	True length (m)	Extension (m)
AF	36·40	0·041
BF	35·00	0·033
CF	46·37	−0·057
CG	54·77	0·011
DG	57·88	−0·028
FG	60·42	0·069

Fig. 42

the drawing, and the true lengths and extensions of the bars are tabulated. There is no necessity for the units of extension and length to be consistent one with another: if they are not, the joint displacements will be given in the same units as the extensions. Equations (23) for the six bars of the framework are as follows:

$$36 \cdot 40 \times 0 \cdot 041 = -5(u_F - u_A) + 20(v_F - v_A) + 30(w_F - w_A)$$
$$35 \cdot 00 \times 0 \cdot 033 = 15(u_F - u_B) - 10(v_F - v_B) + 30(w_F - w_B)$$
$$-46 \cdot 37 \times 0 \cdot 057 = -25(u_F - u_C) - 25(v_F - v_C) + 30(w_F - w_C)$$
$$54 \cdot 77 \times 0 \cdot 011 = 20(u_G - u_C) + 10(v_G - v_C) + 50(w_G - w_C)$$
$$-57 \cdot 88 \times 0 \cdot 028 = 25(u_G - u_D) - 15(v_G - v_D) + 50(w_G - w_D)$$
$$60 \cdot 42 \times 0 \cdot 069 = 45(u_G - u_F) + 35(v_G - v_F) + 20(w_G - w_F)$$

BRACED FRAMEWORKS

Now A, B, C and D are fixed points, so that the twelve displacements with these letters as suffices are zero. The remaining six displacements can be found by solving the above equations (in two sets of three) as $u_F = 0.073$m, $v_F = 0.060$m, $w_F = 0.022$m, $u_G = 0.162$m, $v_G = 0.121$m, $w_G = -0.077$m.

2.8.3. *Virtual work.* The methods for determining the deformations of frameworks discussed in §2.8.1–2 enable the displacements of all of the joints to be found. If the displacements of only a small number of joints are required, a method based on the concept of

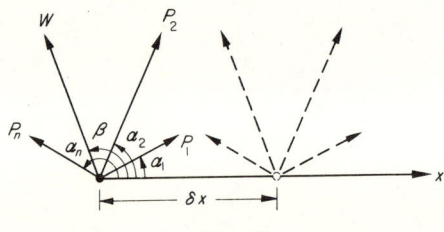

Fig. 43

virtual work may be quicker. This method makes use of one of the properties of a structural system in equilibrium, but it is important to realise that we are using this equilibrium property as a ruse to avoid complicated geometrical reasoning: the real problem with which we are concerned is entirely geometrical and has nothing to do with the equilibrium of the structure.

Consider a joint in a plane truss which is in equilibrium under bar forces $P_1 \ldots P_n$ and external load W applied at angles $\alpha_1 \ldots \alpha_n$ and β to the *x*-direction (Fig. 43). Let the joint suffer a small displacement δx. Since the displacement is small we shall assume that during it the angles and bar forces remain unaltered. Then force P_1 moves in its own line of action by $\delta x \cos \alpha_1$, and similarly for the other forces. The total work done by all of the forces acting on the joint during the displacement δx is

$$P_1 \cdot \delta x \cos \alpha_1 + P_2 \cdot \delta x \cos \alpha_2 + \ldots P_n \cdot \delta x \cos \alpha_n + W \cdot \delta x \cos \beta$$
$$= \delta x (P_1 \cos \alpha_1 + P_2 \cos \alpha_2 + \ldots P_n \cos \alpha_n + W \cos \beta)$$

From equation (11), the term in brackets is zero. It follows that the total work done by the forces acting at a joint of a plane truss during

STATICALLY DETERMINATE TRUSSES

a small displacement from a position of equilibrium is zero. The principle can be generalised to three-dimensional frames and by summation to include all the joints of the structure: it may then be stated in the following form. If we have a framework in equilibrium and we apply to it a set of arbitrary small displacements, the total work done by the internal and external forces acting on the joints will be zero. This statement is a particular form of a general theorem

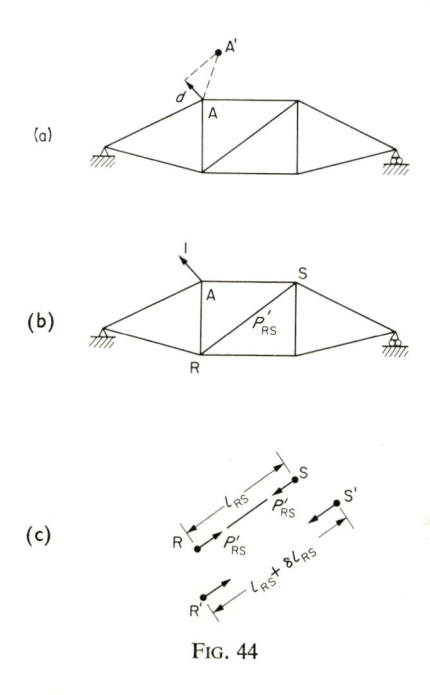

FIG. 44

known as the principle of virtual work: in the form given here it is known as the principle of virtual displacements. We shall now consider how it may be applied to solve the geometrical problem of finding the displacement of a joint of a framework.

Consider the framework shown in Fig. 44(a) and suppose that we wish to find the component of displacement d of joint A in the direction shown, when the members elongate by specified amounts, and A moves to A'. To the undeformed frame, apply a very small load at A in the direction of d (Fig. 44(b)). This load will be known as the unit load (it may be imagined as 10^{-6}N) and it produces no

65

BRACED FRAMEWORKS

measurable deformation of the frame. The forces in the bars of the frame due to the unit load can be calculated by the ordinary methods of equilibrium. Let the force in member RS due to the application of the unit load be P'_{RS}. Now if we apply any arbitrary set of small displacements to the frame, which is in equilibrium under the unit load, the total work done on the joints will be zero, by the principle of virtual work. Let us choose that particular set of displacements which is due to the actual elongations of the bars. Suppose bar RS elongates by δl_{RS}. Then the two forces P'_{RS} acting on joints R and S move in their own line of actions by a total of $-\delta l_{RS}$ and do work $-P'_{RS} \cdot \delta l_{RS}$ (Fig. 44(c)). Due to all of the bar forces and bar elongations the total work done is

$$\sum_{\text{bars}} -P'_{RS} \cdot \delta l_{RS}$$

The only external load which moves in its own line of action is the unit load (the reactive forces at the supports do not move in their own lines of action). It does work $1 \cdot d$. We therefore have

$$1 \cdot d - \sum_{\text{bars}} P'_{RS} \cdot \delta l_{RS} = 0$$

or
$$d = \sum_{\text{bars}} P'_{RS} \cdot \delta l_{RS} \qquad (24)$$

Equation (24) provides a rapid means of finding the component displacement in a specified direction of a single joint of a frame. The equation is based upon the principle of virtual work for a system in equilibrium, but it must be emphasised again that the problem we have solved—that of finding the displacement of a joint due to a given set of bar elongations—is entirely geometrical, and that the equilibrium concerned (that of the frame under the infinitesimal unit load) has no connection with the equilibrium of the frame under a genuine load system.

An important point which may be noted from equation (24) is that the displacement of any point in the structure is a linear function of the extensions of the bars. The unit-load method described here is similar to that of Maxwell (1864b) and Mohr (1874).

We illustrate the use of equation (24) by finding the horizontal displacement of joint F in the plane truss shown in Fig. 45(a). Figure 45(b) shows the unit load and the corresponding forces in the

STATICALLY DETERMINATE TRUSSES

Member	Elongation (cm)
AD	0·17
BD	0·14
CD	−0·33
DF	0·25
CF	−0·41

Fig. 45

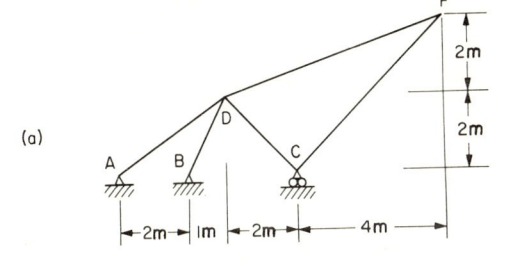

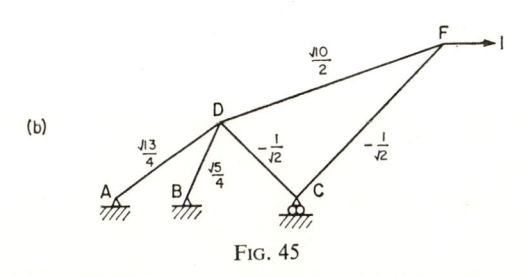

bars. From equation (24) the horizontal displacement of F due to the elongations of the bars shown in the table of Fig. 45(a) is

$$d = \sum_{\text{bars}} P'_{RS} \cdot \delta l_{RS}$$

$$= \frac{\sqrt{13}}{4} \times 0.17 + \frac{\sqrt{5}}{4} \times 0.14 - \frac{1}{\sqrt{2}} \times (-0.33) +$$

$$+ \frac{\sqrt{10}}{2} \times 0.25 - \frac{1}{\sqrt{2}} \times (-0.41)$$

$$= 1.15 \text{ cm}$$

For simplicity, we have assumed in the discussion above that the reactive forces at the supports do not move in their own line of action. If in fact the supports deflect when there is a change of environment, we have additional terms on the right-hand side of equation (24) of

$$\sum_{\text{supports}} (H' . \delta h + V' . \delta v)$$

where H' and V' are the horizontal and vertical components of reactive force applied to the support due to the unit load and δh and δv are the movements of the support in the directions of H' and V'.

2.9. Hooke's Law and the Superposition of Displacements

In §2.6 we investigated the conditions under which it was possible to find the bar forces in a truss by superposing the effects of a number of load systems. We shall now discuss the corresponding problem for displacements.

Consider a statically determinate truss subjected to a single load W at joint A and suppose that we measure the deflection d in a particular direction at joint J (Fig. 46). Then, since the truss is

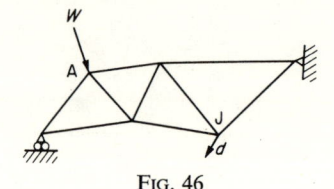

Fig. 46

statically determinate, the bar forces will be proportional to W. Suppose that each bar behaves in a linear elastic manner, so that its extension is proportional to the force applied to it: then the extensions of the bars will be proportional to W. It follows that if we draw a series of displacement diagrams for the truss, for different values of W, these displacement diagrams will all be similar figures and their size will be proportional to W. The displacement of any point in the structure, and in particular the displacement d, is proportional to the magnitude of the applied load W.

This linear relationship between load and deflection is known as Hooke's law (Hooke, 1678). It can be extended to statically indeterminate structures which are free from pre-stress provided their shape does not change significantly under load and provided each bar behaves linearly: under these circumstances it will be seen that doubling the displacement of every point in the structure will double the bar forces and thus provide a system in equilibrium with twice the applied load. Similar reasoning shows that Hooke's law can be used for structures which are pre-stressed provided the same conditions are satisfied and the deflections of the joints are measured from their positions when the structure is free from external load. If the supports of the structure deflect under load, Hooke's law will still be true provided the relation between deflection and load is linear.

If we have a structure whose shape does not change significantly under load, and in which the behaviour of each of the bars is linear elastic, then the deflection at any point J will be proportional to the magnitude of a single load applied at any other point A. Now let us apply a number of loads at joints A, B, C,... The bar forces due to these loads can be found by superposing the bar forces due to the loads acting individually (see §2.6). Further, since the bars are linear-elastic, the extensions of the bars can be found by superposing the extensions due to the individual loads. Finally, since the displacement at J is a linear function of the bar extensions (see §2.8.3), the displacement of J due to the combined loads can be found by superposing the displacements due to the loads acting individually.

We thus find that for a structure obeying Hooke's law, the displacements due to complicated load systems may be found by superposing the displacements due to simpler systems. A structure for which Hooke's law is valid constitutes a linear system (see §1.5) and displacements due to such causes as change of temperature or humidity can also be superimposed, provided they are linear and do not interact. One of the most important applications of the method of superposition of displacements is in the analysis of symmetrical structures: we shall consider this in the next section.

2.10. Properties of Symmetrical Structures

Very many practical structures, because of the type of loading to which they are subjected or for reasons of economics or aesthetics, are made symmetrical about an axis. Examples of symmetrical plane

BRACED FRAMEWORKS

frameworks are shown in Fig. 47. For true symmetry it is necessary that the geometry of the structure should be symmetrical, that the structure should be symmetrically supported, that the cross-sectional areas and material properties of corresponding bars should be identical, and that in the case of a redundant structure the pre-stress in corresponding bars should be equal.

If a symmetrical structure is subjected to symmetrical changes of environment, such as symmetrical loading or symmetrical changes of

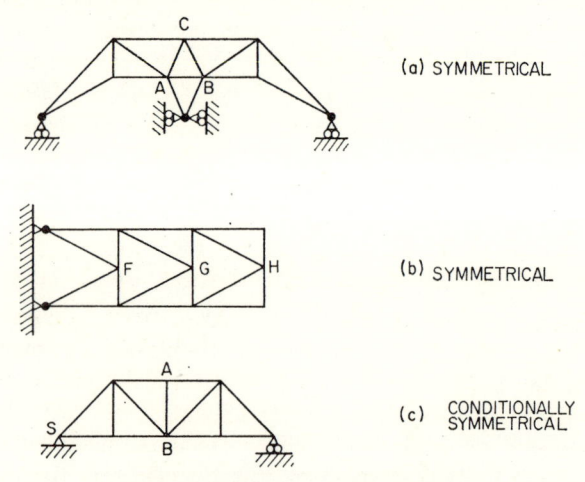

(a) SYMMETRICAL

(b) SYMMETRICAL

(c) CONDITIONALLY SYMMETRICAL

The cross-sectional areas and material properties of bars in corresponding positions are identical

FIG. 47. Symmetrical frameworks

temperature, certain features of the behaviour of the structure may be noted which assist its analysis. First, forces in corresponding bars of the structure are equal. Second, the displacements in the direction of the axis of symmetry of corresponding points are equal. Perpendicular to the axis of symmetry they are of equal magnitude but opposite sign. It follows that points on the axis of symmetry of the structure (such as C in Fig. 47(a) and F, G and H in Fig. 47(b)) are not displaced perpendicular to the axis and central members of the structure (such as AB in Fig. 47(a)) do not rotate. All of these properties may be demonstrated by the same method. Suppose that the force in CA is not equal to the force in CB when the structure

STATICALLY DETERMINATE TRUSSES

shown in Fig. 47(a) is symmetrically loaded. Consider the mirror image of the solution obtained. The structure is unchanged and the loading is unchanged, but the bar forces are different. We thus have two different solutions to the same problem. The only distribution of bar forces which will lead to the same solution in both cases is for the forces in corresponding bars to be equal. Similar arguments can be used for the remaining properties: it is suggested that the reader should develop these for himself, drawing appropriate diagrams.

Figure 47(c) shows a structure which is not symmetrical (since the supports differ) but which can be treated as though it were symmetrical under certain conditions. If the structure is symmetrically loaded there will be no net horizontal force and the reactions at the supports will be vertical and equal. The forces in corresponding bars will be equal and the deflections of corresponding points will be equal except for a horizontal rigid-body displacement (due to the fact that horizontal movement is prevented at S and not at A or B). A rigid-body displacement of the truss (without rotation) does not change the displacement diagram. AB remains vertical.

We illustrate the use of the symmetry properties by analysing the truss shown in Fig. 48(a). This truss is conditionally symmetrical. Under the loading shown there will be a vertical reaction at each support of 50kN. The bar forces are tabulated and the extensions are calculated in the table of Fig. 48(b). (It will be remembered from §1.5 that the longitudinal stress in a bar is P/A, that the strain is P/AE, and the extension thus Pl/AE.) To draw the displacement diagram we make use of the symmetry property that AB does not rotate. a therefore lies on the same vertical line as b on the displacement diagram. Starting at a, we can locate b and then proceed from joint to joint towards one of the supports. The displacement diagram for the right-hand half of the truss is shown in full line in Fig. 48(c). The displacement diagram for the left-hand half of the truss is the mirror image about ab of that for the right-hand half (broken lines). In obtaining the absolute displacements of the truss from Fig. 48(c) it must be remembered that the fixed point is g'.

Two further useful properties of a symmetrical structure are obtained when the structure is subjected to skew-symmetrical changes of environment. For skew-symmetry, the environmental changes at corresponding points of the structure are equal in magnitude but opposite in sign.

BRACED FRAMEWORKS

First, when a statically determinate symmetrical structure is subjected to skew-symmetrical loading, the forces in corresponding bars are equal in magnitude but opposite in sign. Consider the symmetrical truss shown in Fig. 49(a). Let the truss be loaded skew-

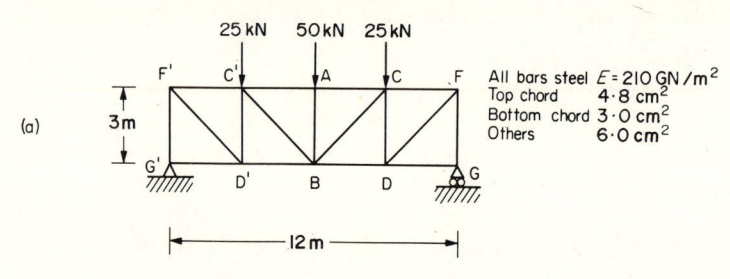

(a)

All bars steel $E = 210 \text{ GN/m}^2$
Top chord $4 \cdot 8 \text{ cm}^2$
Bottom chord $3 \cdot 0 \text{ cm}^2$
Others $6 \cdot 0 \text{ cm}^2$

(b)

Member	Force (P) (kN)	Length (l) (m)	Area (A) (cm^2)	Extension (Pl/AE) (cm)
AB	−50	3	6·0	−0·119
AC, AC′	−75	3	4·8	−0·223
CF, C′F′	−50	3	4·8	−0·149
BD, BD′	50	3	3·0	0·238
DG, D′G′	0	3	3·0	0
CD, C′D′	−50	3	6·0	−0·119
FG, F′G′	−50	3	6·0	−0·119
BC, BC′	25√2	3√2	6·0	0·119
DF, D′F′	50√2	3√2	6·0	0·238

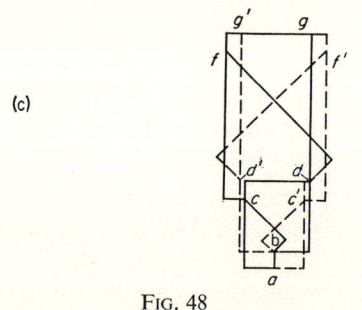

(c)

Fig. 48

symmetrically and let the forces in two corresponding bars be P and Q. Now consider the mirror-image solution (Fig. 49(b)). Since the truss is statically determinate, the bar forces due to two different load systems can be superimposed. The result of superposing the

STATICALLY DETERMINATE TRUSSES

systems of Figs. 49(a) and (b) is shown in Fig. 49(c). The loads are now zero, and the force in each bar is $P+Q$. But in an unloaded statically determinate structure the bar forces are zero. It follows that $P+Q=0$ and $Q=-P$. As a corollary we may note that the force in a centre member such as AB in Fig. 47(a) is zero. The same results are also true for a statically indeterminate structure provided that the behaviour of the members in tension and compression is identical and that any pre-stress is skew-symmetrical.

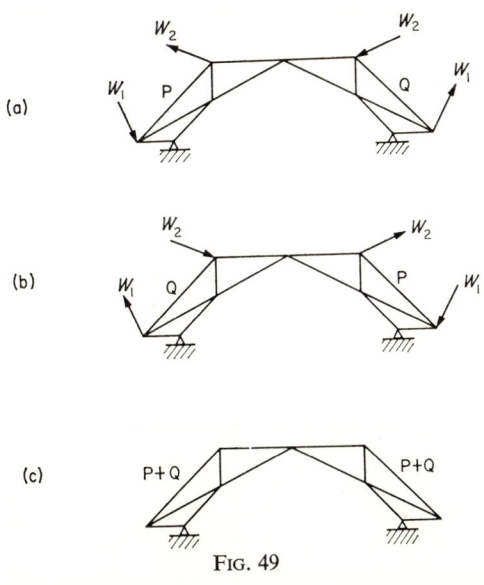

FIG. 49

Secondly, if the load–deformation relationship for corresponding bars is the same in tension as in compression, and in the case of redundant frameworks any pre-stress is skew-symmetrical, then the deflection in the direction of the axis of symmetry of corresponding points will be skew-symmetrical, and perpendicular to the axis of symmetry will be equal. As a particular case we may note that the deflection along the axis of points on the axis of symmetry will be zero. The reader should prove this second property for himself.

A framework such as that shown in Fig. 47(c) may be treated as symmetrical under skew-symmetrical loading provided there is no net horizontal force. If this condition is satisfied the reactions at the supports will be vertical forces of equal magnitude but opposite sign,

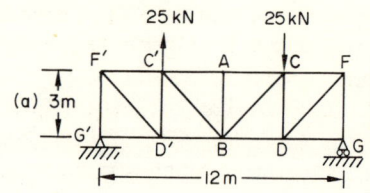

BRACED FRAMEWORKS

(a) [Figure: Truss diagram with 25 kN loads at C and C', spans labeled F'-C'-A-C-F on top chord, G'-D'-B-D-G on bottom chord, 3 m height, 12 m length. All bars steel, $E = 210 \text{ GN/m}^2$. Top chord, 4·8 cm². Bottom chord, 3·0 cm². Others, 6·0 cm².]

(b)

Member	Force (P) (kN)	Length (l) (m)	Area (A) (cm²)	Extension (Pl/AE) (cm)
AB	0	3	6·0	0
AC	0	3	4·8	0
CF	−12·5	3	4·8	−0·037
BD	12·5	3	3·0	0·060
DG	0	3	3·0	0
CD	−12·5	3	6·0	−0·030
FG	−12·5	3	6·0	−0·030
BC	−12·5√2	3√2	6·0	−0·060
DF	12·5√2	3√2	6·0	0·060

(c) [Williot diagram showing displacement 0·265 cm with points a, b, c, d,g, f]

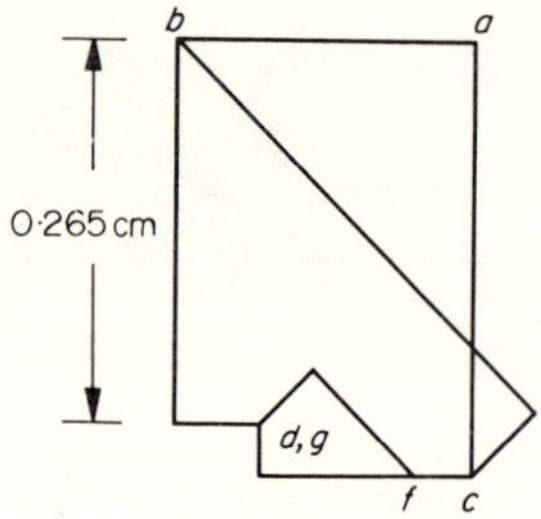

(d) [Williot-Mohr diagram with points b, a, f, c, d, g,g', c', d']

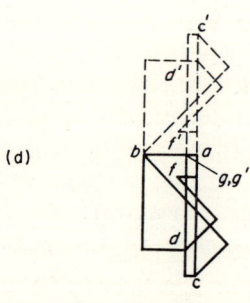

Fig. 50

STATICALLY DETERMINATE TRUSSES

and both of the properties given above will hold. In particular we may note that the horizontal displacement of G will be equal to that of G' and will therefore be zero.

As an example of the use of these further properties, consider the truss of Fig. 48(a) under skew-symmetrical loading, as shown in Fig. 50(a). The forces in and extensions of the bars in the right-hand half of the truss are shown in the table of Fig. 50(b): those for the bars in the left-hand half are of equal magnitude but opposite sign. The Williot diagram for the right-hand half of the truss, starting at G and assuming GD remains horizontal, is shown in Fig. 50(c). This diagram is incorrect, since A and B are found to be 0·265cm above G, whereas points on the axis of symmetry should have no vertical

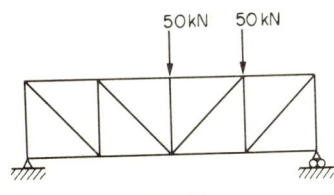

Fig. 51

deflection. The corrected Williot diagram (obtained by adding a rigid-body rotation of 0·265/600 rad anticlockwise) is shown in the full-line diagram of Fig. 50(d). The displacement diagram for the left-hand half of the truss is the mirror image about ab of that for the right-hand half, since this ensures that horizontal displacements of corresponding points are equal and vertical displacements are reversed: it is shown by the broken-line diagram of Fig. 50(d). It is of interest to note that the second half of the displacement diagram is obtained as the mirror image of the first half about points on the axis of symmetry whether the loading be symmetrical or skew-symmetrical.

Under some circumstances the kinds of analysis described above may be useful even though the structure is loaded neither symmetrically nor skew-symmetrically. Any set of environmental changes applied to a symmetrical structure can be divided into symmetrical and skew-symmetrical components. For example, the loading shown in Fig. 51 is compounded of those of Figs. 48(a) and 50(a). More generally, if we have loads W and W' applied at corresponding points A and A' of a symmetrical structure, then these

75

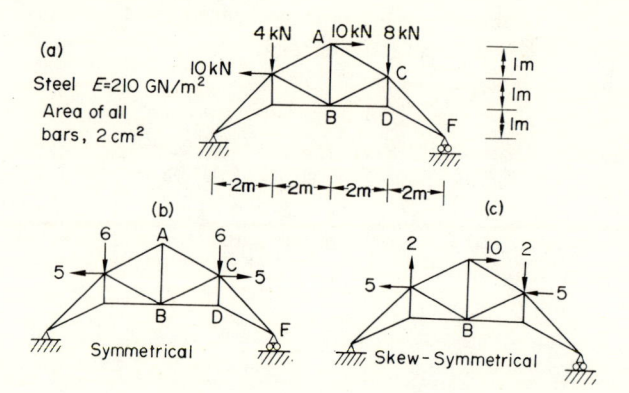

BRACED FRAMEWORKS

(a) Steel $E = 210$ GN/m² Area of all bars, 2 cm²

(b) Symmetrical

(c) Skew-Symmetrical

(d)

Member	Force due to (b) (kN)	Length (m)	Area (cm²)	Elongation (cm)
AC	$-\tfrac{7}{4}\sqrt{5}$	$\sqrt{5}$	2	−0·021
CF	$-12\sqrt{2}$	$2\sqrt{2}$	2	−0·114
BD	12	2	2	0·057
DF	$6\sqrt{5}$	$\sqrt{5}$	2	0·071
AB	$\tfrac{7}{4}$	2	2	0·008
BC	$-\tfrac{7}{4}\sqrt{5}$	$\sqrt{5}$	2	−0·021
CD	6	1	2	0·014

(e) 0·53 cm

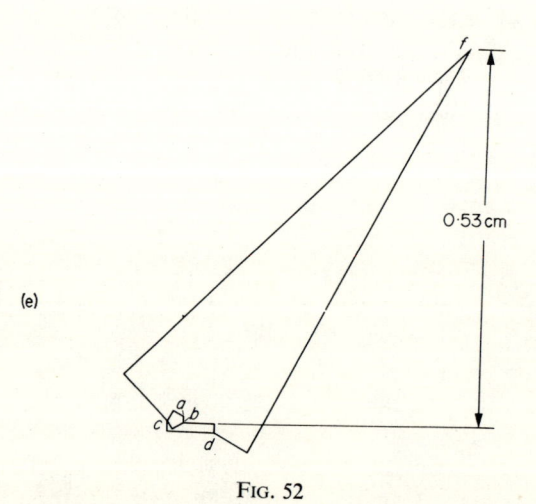

FIG. 52

STATICALLY DETERMINATE TRUSSES

may be replaced by a symmetrical component $(W+W')/2$ applied at both A and A', together with a skew-symmetrical component consisting of $(W-W')/2$ applied at A and $-(W-W')/2$ applied at A'. If the structure obeys Hooke's law, so that the principle of superposition is valid, we can superpose the displacements due to the symmetrical and skew-symmetrical components in order to find the displacements due to the original system. The same method can be used for temperature effects, etc., provided the structural system is linear.

As an example, consider the truss shown in Fig. 52(a) and suppose it is desired to find the vertical deflection of joint B under the given loading. The symmetrical and skew-symmetrical components of the loading are shown in Fig. 52(b) and (c). The elongations of the bars due to the symmetrical component of the loading are calculated in the table of Fig. 52(d). The displacement diagram can be drawn without difficulty, since under symmetrical loading AB does not rotate. It is shown in Fig. 52(e): the vertical deflection of B is equal to 0·53cm. When the truss is subjected to the skew-symmetrical loading shown in Fig. 52(c), points on the axis of symmetry (including B) do not deflect vertically. Superposing the deflections due to loading systems (b) and (c), we find that the vertical deflection of B due to the loads shown in Fig. 52(a) is $0.53+0 = 0.53$cm.

2.11. Maxwell's Reciprocal Theorem

We now proceed to develop a theorem which is of great value in both the theoretical and experimental analysis of structures which obey Hooke's law. Let a load W_{A1} be applied to such a structure in a certain direction at joint A (Fig. 53(a)) and suppose that we wish to find the component of deflection in a certain direction at joint B. We shall denote this deflection by $W_{A1}\delta_{AB}$. δ_{AB} is known as an influence coefficient: it measures the deflection at B per unit load at A. Suppose that the force in a typical bar due to the load W_{A1} is $W_{A1}P'_A$, where P'_A is the force in the bar due to a unit load at A. Let the elongation of the bar due to the force $W_{A1}P'_A$ be $kW_{A1}P'_A$, where k is a constant.

We can find the deflection at B by the method of virtual work (§2.8.3). Apply a unit load at B (Fig. 53(b)). Let the force in a

77

BRACED FRAMEWORKS

typical bar due to the unit load at B be P'_B. Then the deflection at B due to the load W_{A1} at A is given by equation (24) as

$$W_{A1}\delta_{AB} = \sum_{\text{bars}} P'_B . kW_{A1} P'_A \qquad (25)$$

Now suppose that a load W_{B2} is applied to the structure at B in the same direction as that in which the displacement was measured

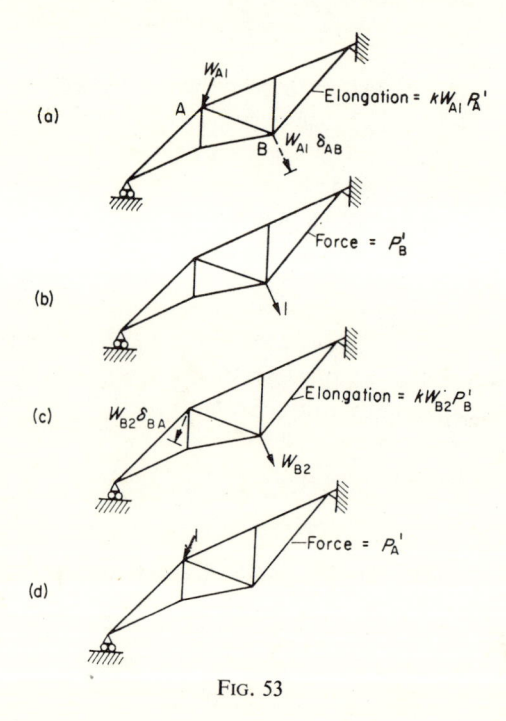

Fig. 53

(Fig. 53(c)) and that we wish to find the component of deflection at A in the same direction as that in which load W_A was applied. We shall denote this deflection by $W_{B2}\delta_{BA}$. The force in a typical bar due to W_{B2} is $W_{B2}P'_B$ and the elongation of the bar is $kW_{B2}P'_B$. If we apply a unit load at A (Fig. 53(d)), the force in a typical bar is P'_A. Using equation (24) again we find that the deflection at A due to the load W_{B2} at B is given by

$$W_{B2}\delta_{BA} = \sum_{\text{bars}} P'_A . kW_{B2} P'_B \qquad (26)$$

STATICALLY DETERMINATE TRUSSES

If we multiply equation (25) by W_{B2} and equation (26) by W_{A1}, the right-hand sides become equal. It follows that

$$W_{A1}(W_{B2}\delta_{BA}) = W_{B2}(W_{A1}\delta_{AB}) \qquad (27)$$

This is Maxwell's reciprocal theorem (Maxwell, 1864b). It can be generalised in the following form, due to Betti (1872). Suppose that a set of loads $W_{A1}, W_{B1}, W_{C1}, \ldots$, act in certain directions at points $A, B, C, \ldots$, of the structure (some of the loads can be zero). Let the displacements at $A, B, C, \ldots$, in the directions of the applied loads be $\Delta_{A1}, \Delta_{B1}, \Delta_{C1}, \ldots$ Now let a second set of loads $W_{A2}, W_{B2}, W_{C2}, \ldots$, be applied at $A, B, C, \ldots$, in the same directions as the loads of the first set, and let the displacements at $A, B, C, \ldots$, in the directions of the loads be $\Delta_{A2}, \Delta_{B2}, \Delta_{C2}, \ldots$ Then

$$W_{A1} \cdot \Delta_{A2} + W_{B1} \cdot \Delta_{B2} + W_{C1} \cdot \Delta_{C2} \ldots$$
$$= W_{A2} \cdot \Delta_{A1} + W_{B2} \cdot \Delta_{B1} + W_{C2} \cdot \Delta_{C1} \ldots \qquad (28)$$

This general form of Maxwell's reciprocal theorem can be obtained by superposition from equation (27) if we remember that the deflections are given by equations of the type

$$\Delta_{A1} = W_{A1} \cdot \delta_{AA} + W_{B1} \cdot \delta_{BA} + W_{C1} \cdot \delta_{CA} \ldots$$

Equation (28) reduces to equation (27) if all loads except W_{A1} and W_{B2} are zero.

If equation (27) is divided by $W_{A1} W_{B2}$ we obtain

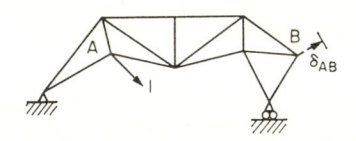

$$\delta_{BA} = \delta_{AB} \qquad (29)$$

This is the most generally useful form of the reciprocal theorem. It

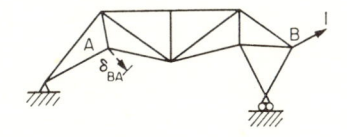

FIG. 54. Maxwell's reciprocal theorem

BRACED FRAMEWORKS

Light alloy, $E = 70\,\text{GN/m}^2$
Chords 60 cm^2
Others 30 cm^2
6 m
36 m

(a)

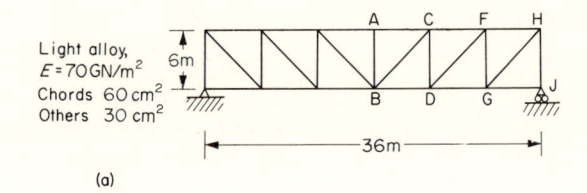

(b)

Member	Force due to 1 kN at B (kN)	Length (m)	Area (cm^2)	Elongation (cm)
AC	$-\dfrac{3}{2}$	6	60	−0·00214
CF	−1	6	60	−0·00143
FH	$-\dfrac{1}{2}$	6	60	−0·00071
BD	1	6	60	0·00143
DG	$\dfrac{1}{2}$	6	60	0·00071
GJ	0	6	60	0
AB	0	6	30	0
CD	$-\dfrac{1}{2}$	6	30	−0·00143
FG	$-\dfrac{1}{2}$	6	30	−0·00143
HJ	$-\dfrac{1}{2}$	6	30	−0·00143
BC	$\dfrac{1}{\sqrt{2}}$	$6\sqrt{2}$	30	0·00286
DF	$\dfrac{1}{\sqrt{2}}$	$6\sqrt{2}$	30	0·00286
GH	$\dfrac{1}{\sqrt{2}}$	$6\sqrt{2}$	30	0·00286

Fig. 55(a), (b)

STATICALLY DETERMINATE TRUSSES

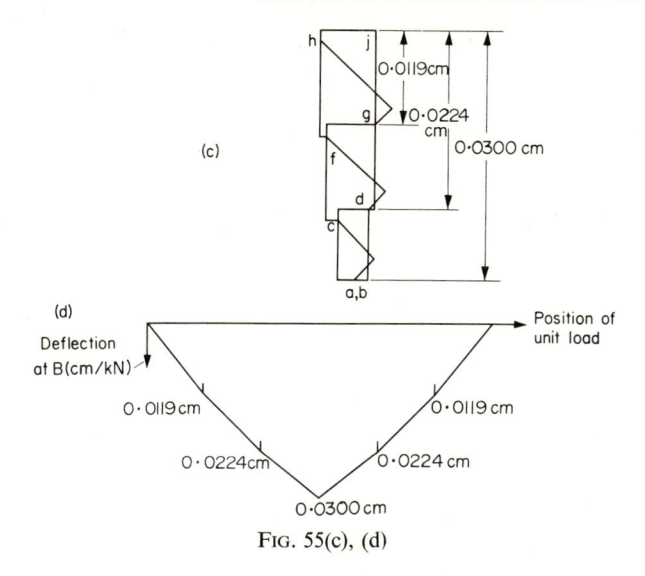

Fig. 55(c), (d)

states that if we apply a unit load in a certain direction at point A in a structure, and measure the component deflection in a certain direction at point B, the result will be the same as if we applied the unit load at B and measured the deflection at A. This is illustrated in Fig. 54. Lord Rayleigh (1873) extended the reciprocal theorem to vibrating systems.

As an example of the use of Maxwell's reciprocal theorem we shall consider the problem of finding the influence line for the vertical deflection of joint B of the truss shown in Fig. 55(a) when unit load crosses the lower chord. A direct solution would require us to place unit load at each of B, D and G in turn, calculate the forces in and elongations of the bars, and draw displacement diagrams to find the deflection of B. This tedious procedure can be avoided by the use of Maxwell's reciprocal theorem. The deflection of B due to a unit load at B, D or G is the same as the deflection at B, D or G due to a unit load at B. We therefore consider the truss subjected to a unit load at B. The elongations of the members are given in the table of Fig. 55(b) and the displacement diagram in Fig. 55(c) (since the truss is a symmetrical structure symmetrically loaded, AB remains vertical). The deflections at B, D and G due to the unit load at B are 0·0300, 0·0224 and 0·0119cm/kN. The deflections at B due to unit loads at

BRACED FRAMEWORKS

B, D or G will have the same values. The influence line is plotted in Fig. 55(d).

Maxwell's reciprocal theorem is used in a number of experimental methods for the analysis of model structures, notably those due to Beggs (1927). It must be emphasised that the theorem is only applicable to structures (statically determinate or redundant) which obey Hooke's law. If Hooke's law is not obeyed, the reciprocal theorem is invalid.

EXAMPLES

2(a). Determine the forces in the members of the truss shown in Fig. 56 by the method of resolving at the joints.

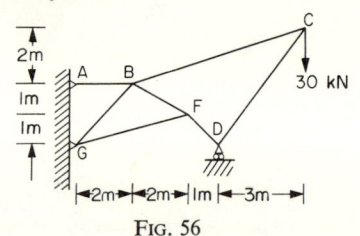

Fig. 56

2(b). Determine the forces in the members of the truss shown in Fig. 56 by using Bow's notation to draw a Maxwell diagram.

2(c). Use the method of sections to find the force in bar AB of the truss shown in Fig. 57.

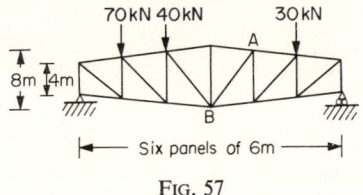

Fig. 57

2(d). Find the forces in the members of the truss shown in Fig. 58. (*Hint:* Use Henneberg's method and replace AF by CH.)

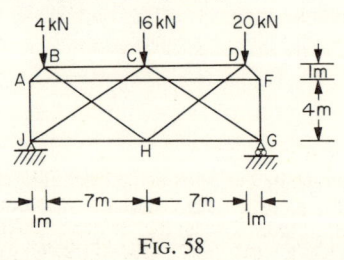

Fig. 58

STATICALLY DETERMINATE TRUSSES

2(e). Use the method of tension coefficients to find the forces in the bars of the framework shown in Fig. 59. (*Hint:* Make use of symmetry.) Check the forces in AF and BG by the method of sections.

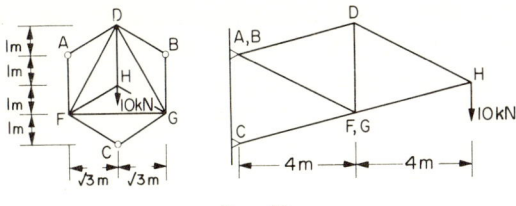

FIG. 59

2(f). Determine which of the frameworks shown in Fig. 60 are statically determinate.

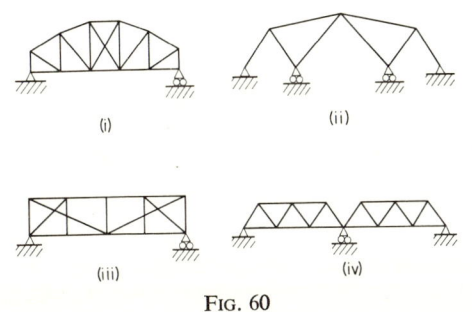

FIG. 60

2(g). Draw an influence line for the force in member AB of the truss shown in Fig. 61(a) as unit load crosses the lower chord. Find the maximum force in AB due to the train of loads shown in Fig. 61(b). The loads cannot turn round.

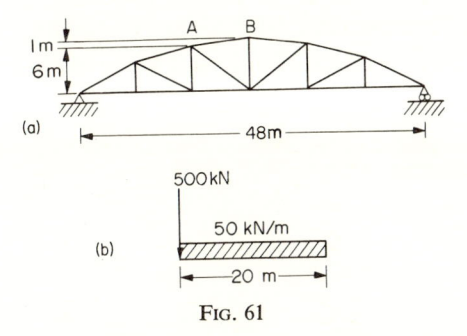

FIG. 61

2(h). The truss shown in Fig. 56 is made from light alloy having a coefficient of expansion of $22 \times 10^{-6}/°C$. By drawing a displacement diagram find the vertical displacement of joint C when the members suffer the following temperature increases: AB, 100°C; BC, 10°C; GB, BF, 20°C; GF, FD, DC, 50°C.

BRACED FRAMEWORKS

2(i). Solve question 2(h) by the algebraic method of §2.8.2.

2(j). Solve question 2(h) by the method of virtual work.

2(k). Find the displacement of each of the joints of the truss shown in Fig. 62 due to the given load.

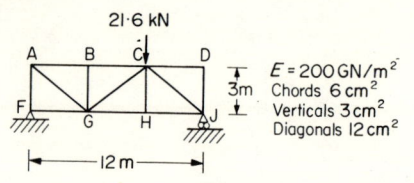

```
                    21·6 kN
                       ↓
       A     B     C     D        ┬    E = 200 GN/m²
                                  3m   Chords 6 cm²
       F                          ┴    Verticals 3 cm²
       ▨    G     H     J              Diagonals 12 cm²
                                ▨
       |─────── 12 m ───────|
```

FIG. 62

2(l) Find the vertical displacement of joint F of the truss shown in Fig. 63, (i) when $W_1 = W_2 = 60$kN, (ii) when $W_1 = -W_2 = 60$kN, and (iii) when $W_1 = 120$kN, $W_2 = 0$. Why is the answer to (iii) not equal to the sum of those to (i) and (ii)?

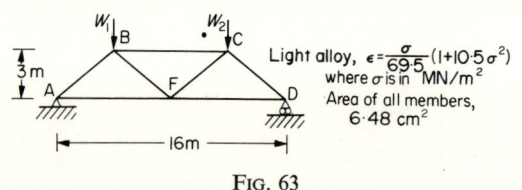

```
              W₁↓B      W₂↓C
         ┬                              Light alloy, ε = σ/69·5 (1+10·5σ²)
         3m   A    F    D               where σ is in MN/m²
         ┴   ▨              ▨           Area of all members,
             |────── 16 m ──────|       6·48 cm²
```

FIG. 63

2(m). In the framework shown in Fig. 64, all of the members have the same cross-sectional area A and Young's modulus E. Find the deflection at the point of application of the load P.

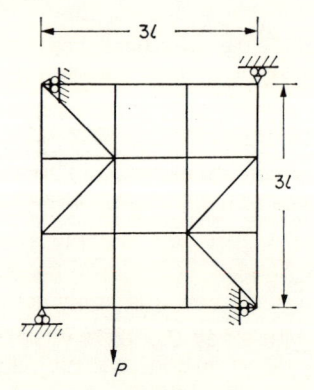

FIG. 64

2(n). The truss shown in Fig. 59 is subjected to a vertical load of WN at D. Find the horizontal movement of H if bars DF, FG and GD have the same cross-sectional area Am² and Young's modulus EN/m². (*Hint:* Use Maxwell's reciprocal theorem and symmetry.)

3. redundant trusses

3.1. Introduction

When analysing a statically determinate framework we first found the bar forces from considerations of equilibrium (§2.4), then the bar elongations from the environment–deformation relationships (§1.5) and finally the displacements of the joints from geometrical considerations (§2.8). With a statically indeterminate framework this simple order of analysis is no longer possible, since the bar forces depend upon compatibility of deformation as well as upon equilibrium. Provided the frame is well conditioned (see §2.5), however, it is possible to adapt our method for statically determinate trusses to the solution of redundant trusses. Suppose we have a truss with m additional bars or reactions over and above those needed for statical determinacy. Denote the forces in those bars by $R_1 \ldots R_r \ldots R_m$. Remove the additional bars and apply across each of the gaps so formed the appropriate pair of forces R_r. We now have a well-conditioned statically determinate frame subjected to external loads and m pairs of forces $R_1 \ldots R_m$. Since the angles between the bars are not significantly dependent on the bar forces, we can find the force in each bar as a linear function of $R_1 \ldots R_m$. For any set of numerical values which we may assign to $R_1 \ldots R_m$, we can find the

BRACED FRAMEWORKS

elongations of the bars and draw a displacement diagram for the statically determinate truss. We can thus find the length of each of the gaps $1\ldots m$: we can also find the length of each of the additional bars $1\ldots m$ when subjected to this set of forces $R_1\ldots R_m$. The successful analysis of a redundant truss depends upon finding that set of values of $R_1\ldots R_m$ such that in each case the length of the member is equal to the length of the gap, and the frame fits together. Much of the present chapter will be concerned with means for solving this problem. In §3.3 the lengths of gap and member will be compared directly. In §§3.4 and 3.5 the comparison is indirect. Whatever method may be used, the essential process remains the same: we find the bar forces in algebraic terms, by consideration of equilibrium, and then choose that set of bar forces such that we satisfy the geometrical requirement that the structure shall fit together.

If the frame is ill conditioned, the method described above is no longer possible. In such cases, it is probably best to define the displacements of the joints symbolically, find the elongations of the members and thus the forces in them, and then write down the equilibrium equations at each joint (see Example 3(c)).

Before discussing the analysis of redundant trusses in detail, it will be worthwhile to consider which of the methods and theorems developed in Chapter 2 may still be applicable to statically indeterminate structures. All of the methods for the determination of bar forces discussed in §2.4 can still be used, although in algebraic terms instead of numerically. The determination of the final displacements of a redundant structure is identical with that for a statically determinate one, since once the bar forces are known, the additional members can be omitted and the displacements found by considering the statically determinate structure which remains: the methods of §2.8 are thus directly applicable. The principle of superposition can be used to determine the bar forces and displacements due to the sum of a number of load systems provided the structure obeys Hooke's law and is free from pre-stress. The concept of influence lines for forces or deflections is applicable to redundant structures, provided the principle of superposition is valid. The properties of symmetrical structures discussed in §2.10 are still true for statically indeterminate structures, under the conditions stated in that section. Maxwell's reciprocal theorem is applicable to any structure (statically determinate or redundant) which obeys Hooke's law.

3.2. Number and Choice of Redundancies

The number of redundant bar forces or reactive forces in a framework can be found by sketching the construction of the structure or alternatively by a modification of equations (19) and (20). Suppose that a plane truss has b bar forces, r components of reactive force and j joints. Then the number of forces over and above those needed for statical determinacy is

$$m = b + r - 2j \qquad (30)$$

The corresponding equation for a three-dimensional frame is

$$m = b + r - 3j \qquad (31)$$

Figures 65(a), (b) and (c) show examples of redundant structures. The assembly shown in Fig. 65(d) has $m = 1$, but it is a mechanism: the bars are not correctly arranged.

There is not generally any particular bar or number of bars in a framework which must be regarded as redundant. Usually a wide

(a) $b = 11$, $r = 4$, $j = 6$, $m = 3$ REDUNDANT STRUCTURE

(b) $b = 16$, $r = 3$, $j = 9$, $m = 1$ REDUNDANT STRUCTURE

(c) $b = 29$, $r = 8$, $j = 16$, $m = 5$ REDUNDANT STRUCTURE

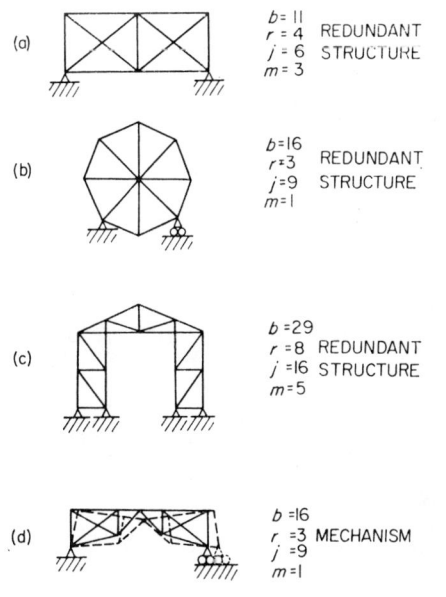

(d) $b = 16$, $r = 3$, $j = 9$, $m = 1$ MECHANISM

FIG. 65. Redundant frameworks

BRACED FRAMEWORKS

choice is possible. Figures 66(a) and (b) show possible choices for the three redundancies in the truss of Fig. 65(a). The choice is not unlimited. The assembly of bars left when the redundant bars or supports have been removed must be statically determinate. The systems shown in Figs. 66(c) and (d) are not satisfactory, since the first is a mechanism and the second is statically indeterminate.

(a) STATICALLY DETERMINATE

(b) STATICALLY DETERMINATE

(c) MECHANISM

(d) STATICALLY INDETERMINATE

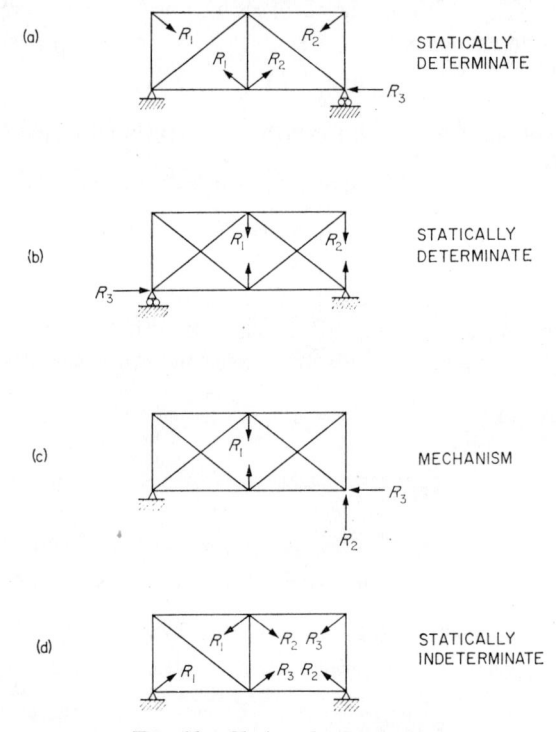

FIG. 66. Choice of reduncancies

The choice of which bar forces or reactive forces to regard as the redundancies will be determined by convenience. If the force in a certain bar is required, it will probably be best to choose this as one of the redundancies, as the force will be known without further calculation once the redundancies have been determined. In many problems the choice is quite unimportant.

3.3. Direct Comparison of Lengths

As already explained in §3.1, the solution of redundant trusses essentially depends on matching the lengths of the members to the gaps in the structure into which they have to fit. A means of solving redundant trusses in this way was discovered independently by Maxwell (1864b) and by Mohr (1874). In the present section we shall compare lengths directly. The size of the gaps in the structure can be found by any of the methods of §2.8.

We begin by considering the frame shown in Fig. 67(a), which has one redundancy. All of the members are made from a material having the stress–strain curve in tension or compression shown in Fig. 67(b). There is no initial lack of fit. Let us choose BC as the redundant bar and denote the force in it by RMN. Then the forces in the remaining bars can be obtained by consideration of equilibrium and are as shown in Fig. 67(c). For any given value of R we can calculate the force in each bar and thence the stress. From the stress–strain curve of Fig. 67(b) we can find the strain in each bar, and on multiplying the strain by the length we obtain the elongation. The elongations of the bars for $R = 24$, 27 and 30MN are shown in the table of Fig. 67(d).

For each value of R we can draw a displacement diagram for the truss of Fig. 67(c). The diagrams for $R = 24$, 27 and 30MN are shown in Fig. 67(e). The distance between b and c, measured in the direction BC (in this case vertically), represents the increase in gap BC due to the application of the 80MN load and the two forces R. Adding this increase to the original length of the gap (50·0m), we find that for $R = 24$, 27 and 30MN, the length of gap BC is 50·70, 50·51 and 50·38m. This relationship is plotted in Fig. 67(f).

For any given value of R we can also find the stress in bar BC, the strain (from Fig. 67(b)), and thence the elongation. Adding the elongation to the original length (50m, since there is no initial lack of fit), we obtain the total length of bar BC as a function of R: this is plotted in Fig. 67(f).

Now, for the solution of the redundant truss to be correct, two conditions have to be satisfied. First, for equilibrium, the force R in bar BC must equal the force R applied to the rest of the frame at B and C. Second, for compatibility, the lengths of gap BC and bar

BRACED FRAMEWORKS

BC must be equal. These conditions are both satisfied at the point where the two curves of Fig. 67(f) intersect. The force R is equal to 27·8MN and the length BC is equal to 50·47m. Once R is known, the remaining bar forces can be calculated from Fig. 67(c). We find $P_{AB} = 29\cdot3$MN, $P_{BD} = -46\cdot3$MN, $P_{AC} = 87\cdot0$MN, $P_{CD} = -55\cdot0$MN. From these forces the true elongations of the bars could be found and a correct displacement diagram drawn.

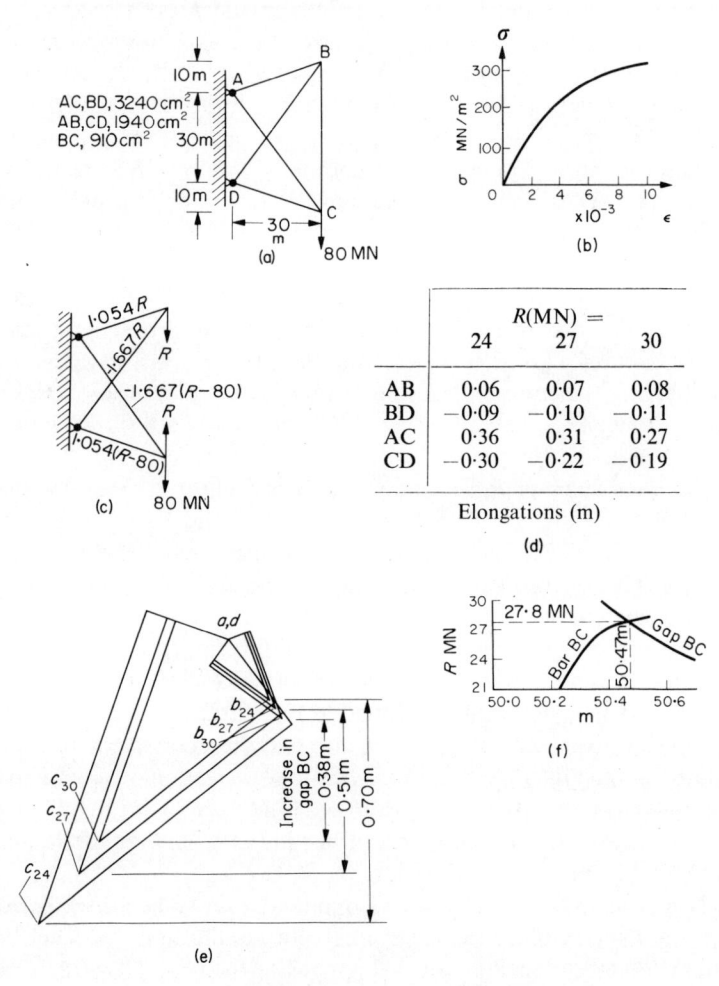

FIG. 67 Direct comparison of lengths

REDUNDANT TRUSSES

In the previous example, the stress–strain relationship was given as an experimental graph and the solution had to be obtained by plotting curves obtained from calculations for a number of values of the unknown force R. If the stress–strain curve can be represented with sufficient accuracy by a simple algebraic expression, the labour of calculation is reduced since a direct solution becomes possible.

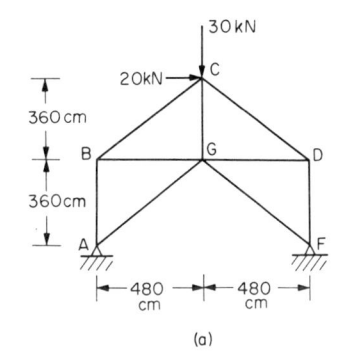

BC, CD, AG, GF $2{\cdot}5\,\text{cm}^2$
AB, CG, DF $3{\cdot}0\,\text{cm}^2$
BG, GD $2{\cdot}0\,\text{cm}^2$
$\epsilon = 2 \times 10^{-4}\sigma + 10^{-6}\sigma^3$, where σ is in kN/cm^2
After stress-free assembly, abutments move so that AF becomes 975 cm

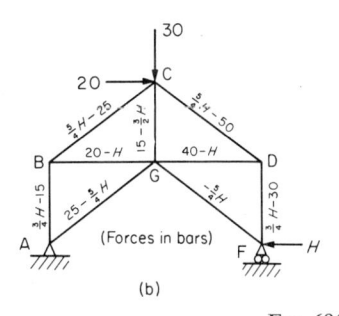

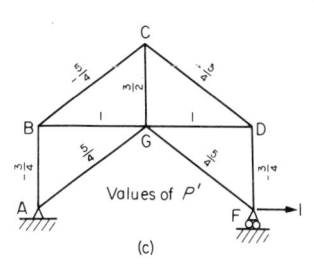

Fig. 68(a)–(c)

Consider the arch shown in Fig. 68(a). The areas of the bars are given in the diagram and the stress–strain relationship for the material can be represented by the equation

$$\varepsilon = 2 \times 10^{-4}\sigma + 10^{-6}\sigma^3 \qquad (32)$$

where σ is in kN/cm^2. The arch is assembled to the dimensions shown in Fig. 68(a) without pre-stress, but subsequently the abutments A and F move so that the distance AF becomes 975 cm. The arch has one redundancy: let us choose this as the horizontal component of

BRACED FRAMEWORKS

Bar	σ (kN/cm^2)	ε	Length (cm)	P'	$P' \times$ elongation (cm)
BC	$\frac{1}{2}H - 10$	$(-30 + \frac{5}{2}H - \frac{3}{40}H^2 + \frac{1}{800}H^3) \times 10^{-4}$	600	$+\frac{5}{4}$	$+2.25 - 0.188H + 0.00563H^2 - 0.000094H^3$
CD	$\frac{1}{2}H - 20$	$(-120 + 7H - \frac{3}{20}H^2 + \frac{1}{800}H^3) \times 10^{-4}$	600	$+\frac{5}{4}$	$+9.00 - 0.525H + 0.00375H^2 - 0.000094H^3$
AG	$-\frac{1}{2}H + 10$	$(30 - \frac{5}{2}H + \frac{3}{40}H^2 - \frac{1}{800}H^3) \times 10^{-4}$	600	$+\frac{5}{4}$	$+2.25 - 0.188H + 0.00563H^2 - 0.000094H^3$
GF	$-\frac{1}{2}H$	$(-H - \frac{1}{800}H^3) \times 10^{-4}$	600	$+\frac{5}{4}$	$-0.075H - 0.000094H^3$
AB	$\frac{1}{4}H - 5$	$(-\frac{45}{4} + \frac{11}{16}H - \frac{3}{320}H^2 + \frac{1}{6400}H^3) \times 10^{-4}$	360	$-\frac{3}{4}$	$+0.30 - 0.019H + 0.00025H^2 - 0.000004H^3$
CG	$-\frac{1}{2}H + 5$	$(\frac{45}{4} - \frac{11}{8}H - \frac{3}{80}H^2 - \frac{1}{800}H^3) \times 10^{-4}$	360	$+\frac{3}{2}$	$+0.61 - 0.074H + 0.00203H^2 - 0.000068H^3$
DF	$\frac{1}{4}H - 10$	$(-30 + \frac{5}{4}H - \frac{3}{160}H^2 + \frac{1}{6400}H^3) \times 10^{-4}$	360	$-\frac{3}{4}$	$+0.81 - 0.034H + 0.00051H^2 - 0.000004H^3$
BG	$-\frac{1}{2}H + 10$	$(30 - \frac{5}{2}H + \frac{3}{40}H^2 - \frac{1}{800}H^3) \times 10^{-4}$	480	$+1$	$+1.44 - 0.120H + 0.00360H^2 - 0.000060H^3$
GD	$-\frac{1}{2}H + 20$	$(120 - 7H + \frac{3}{20}H^2 - \frac{1}{800}H^3) \times 10^{-4}$	480	$+1$	$+5.76 - 0.336H + 0.00720H^2 - 0.000060H^3$
				Total	$+22.42 - 1.559H + 0.0286H^2 - 0.00057H^3$

Fig. 68(d)

REDUNDANT TRUSSES

reactive force at support F and denote it by H kN. Then the forces in the bars are as shown in Fig. 68(b). The corresponding stresses are shown in the second column of the table of Fig. 68(d). The strains, calculated from equation (32), are given in the third column of the table and the lengths of the members in the fourth column. By multiplying these quantities together we could find the elongations of the bars, although in fact we shall not do so.

To find the change in the gap AF in the arch we shall use the method of virtual work (§2.8.3). The forces (P') in the bars due to a unit load applied across AF are shown in Fig. 68(c) and in the fifth column of the table of Fig. 68(d). The final column of the table, which is obtained by multiplying together columns three, four and five, shows for each bar the product of the force (P') of Fig. 68(c) with the elongation due to the load system of Fig. 68(b). From equation (24) the sum of these terms is equal to the increase in gap AF in the arch due to the load system of Fig. 68(b). This total is given in Fig. 68(d). Since the original distance AF was equal to 960cm, the final gap AF in the arch becomes

$$960 + 22 \cdot 42 - 1 \cdot 559 H + 0 \cdot 0286 H^2 - 0 \cdot 00057 H^3$$

For compatibility, the arch must fit between the abutments A and F which are 975cm apart. We therefore have

$$\underbrace{960 + 22 \cdot 42 - 1 \cdot 559 H + 0 \cdot 0286 H^2 - 0 \cdot 00057 H^3}_{\text{(gap AF in arch)}} = \underbrace{975}_{\text{(gap AF between abutments)}}$$

This equation yields $H = 5 \cdot 2$kN. The remaining bar forces can now be found from Fig. 68(b), or alternatively the stresses from the second column of the table in Fig. 68(d). We find $\sigma_{BC} = -7 \cdot 4$, $\sigma_{CD} = -17 \cdot 4$, $\sigma_{AG} = 7 \cdot 4$, $\sigma_{GF} = -2 \cdot 6$, $\sigma_{AB} = -3 \cdot 7$, $\sigma_{CG} = 2 \cdot 4$, $\sigma_{DF} = -8 \cdot 7$, $\sigma_{BG} = 7 \cdot 4$, $\sigma_{GD} = 17 \cdot 4$kN/cm².

If the relationship between stress and strain is linear, a solution of the kind described above becomes particularly simple. The next example concerns a structure which behaves as a linear system. The truss is shown in Fig. 69(a). It is symmetrical and we shall make use of the principle of superposition and of the properties of symmetrical structures. The truss is assembled without pre-stress at 0°C and is then subjected to the temperatures shown. The temperature rises can be divided into symmetrical and skew-symmetrical components as shown in Figs. 69(b) and (c).

93

BRACED FRAMEWORKS

Now if a symmetrical structure is subjected to a skew-symmetrical change of environment, the deflections of corresponding points in directions parallel to the axis of symmetry are of equal magnitude but opposite sign: further, points on the axis of symmetry do not deflect along the axis (see §2.10). It follows that due to the temperature changes shown in Fig. 69(c), F'BF would remain a straight line even if the structure were not connected to one of the supports. The supports therefore impose no restraining forces on the structure, and the bar forces due to the skew symmetrical temperature changes

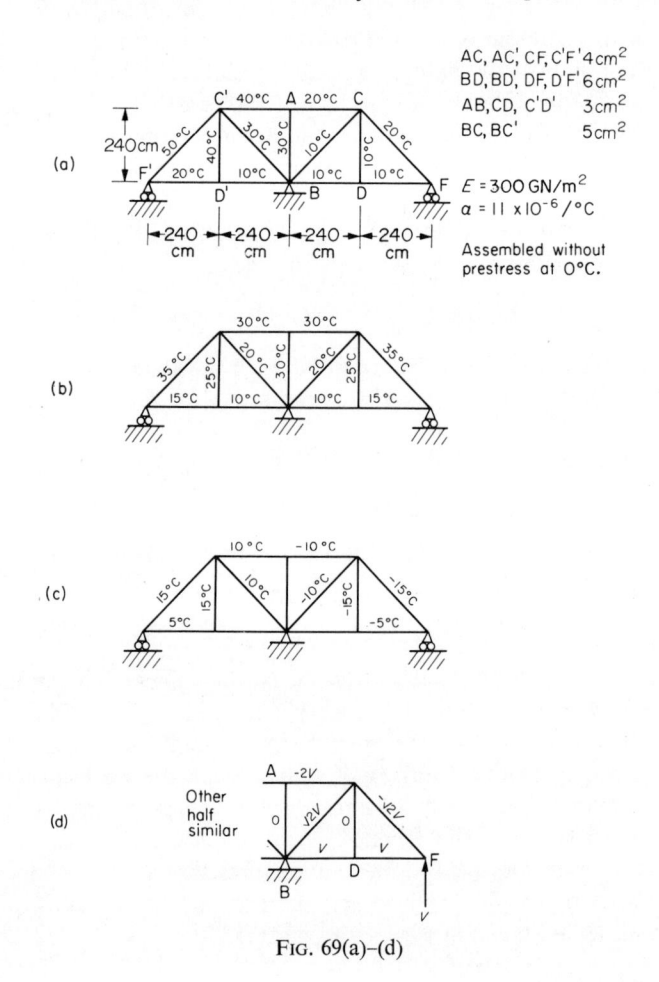

AC, AC', CF, C'F' 4 cm^2
BD, BD', DF, D'F' 6 cm^2
AB, CD, C'D' 3 cm^2
BC, BC' 5 cm^2

$E = 300$ GN/m^2
$\alpha = 11 \times 10^{-6}/°C$

Assembled without prestress at 0°C.

Fig. 69(a)–(d)

REDUNDANT TRUSSES

Bar	Force (N)	Area (cm²)	Stress (N/cm²)	Strain due to stress	Temp. rise (°C)	Strain due to temp.	Total strain	Length (cm)	Elongation (cm)
AC	$-2V$	4	$-\frac{1}{2}V$	$-\frac{1}{60}V \times 10^{-6}$	30	330×10^{-6}	$\left(-\frac{1}{60}V + 330\right) \times 10^{-6}$	240	$(-4\cdot00V + 79200) \times 10^{-6}$
CF	$-\sqrt{2}.V$	4	$-\frac{1}{2\sqrt{2}}V$	$-\frac{1}{60\sqrt{2}}V \times 10^{-6}$	35	385×10^{-6}	$\left(-\frac{1}{60\sqrt{2}}V + 385\right) \times 10^{-6}$	$240\sqrt{2}$	$(-4\cdot00V + 130700) \times 10^{-6}$
BD	V	6	$\frac{1}{6}V$	$\frac{1}{180}V \times 10^{-6}$	10	110×10^{-6}	$\left(\frac{1}{180}V + 110\right) \times 10^{-6}$	240	$(1.33V + 26400) \times 10^{-6}$
DF	V	6	$\frac{1}{6}V$	$\frac{1}{180}V \times 10^{-6}$	15	165×10^{-6}	$\left(\frac{1}{180}V + 165\right) \times 10^{-6}$	240	$(1\cdot33V + 39600) \times 10^{-6}$
AB	0	3	0	0	30	330×10^{-6}	330×10^{-6}	240	79200×10^{-6}
CD	0	3	0	0	25	275×10^{-6}	275×10^{-6}	240	66000×10^{-6}
BC	$\sqrt{2}.V$	5	$\frac{\sqrt{2}}{5}V$	$\frac{\sqrt{2}}{150}V \times 10^{-6}$	20	220×10^{-6}	$\left(\frac{\sqrt{2}}{150}V + 220\right) \times 10^{-6}$	$240\sqrt{2}$	$(3\cdot20V + 74600) \times 10^{-6}$

Fig. 69(e)

BRACED FRAMEWORKS

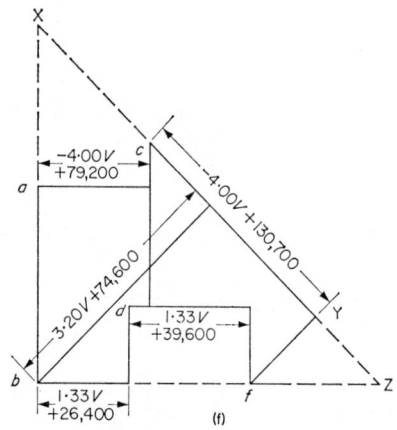

FIG. 69(f). A factor 10^{-6} has been omitted throughout.
Lengths are arbitrary, since V is unknown

are zero. It remains to find the bar forces due to the symmetrical temperature changes of Fig. 69(b).

The truss has one redundancy: let us choose it as the reaction at the right-hand support, V N. The forces in the bars are shown in Fig. 69(d), and the elongations due to these forces and the temperature changes of Fig. 69(b) are calculated in the table of Fig. 69(e). In Fig. 69(f) the displacement diagram for the right-hand half of the truss has been sketched, using the symmetry condition that AB remains vertical. The lengths are arbitrary (since V is not yet known), but the diagram has been drawn so as to satisfy the compatibility condition that b and f are at the same level. The broken lines and points X, Y and Z have been added for the purposes of the ensuing calculation.

We have:
$$Xc = \sqrt{2}(-4 \cdot 00V + 79{,}200)$$
$$XZ = 2(3 \cdot 20V + 74{,}600)$$
Therefore $YZ = XZ - Xc - cY$
$$= 2(3 \cdot 20V + 74{,}600) - \sqrt{2}(-4 \cdot 00V + 79{,}200) -$$
$$-(-4 \cdot 00V + 130{,}700)$$
$$= 16 \cdot 06V - 93{,}500$$
Now $fZ = \sqrt{2}(YZ)$
$$= 22 \cdot 71V - 132{,}200$$

REDUNDANT TRUSSES

Therefore $bZ = bd + df + fZ$
$= 1·33V + 26,400 + 1·33V + 39,600 + 22·71V - 132,200$
$= 25·4V - 66,200$

But $bZ = \sqrt{2}(3·20V + 74,600)$

Therefore $25·4V - 66,200 = \sqrt{2}(3·20V + 74,600)$

whence $V = 8230\text{N}$

Instead of sketching the displacement diagram we could have used the method of virtual work to find the vertical displacement of F and then equated this to zero. The forces in the bars due to a unit vertical load at F are given in the second column of the Table of Fig. 69(e) provided the factor V is omitted. Multiplying these forces by the elongations in the final column of the table and summing we obtain the vertical deflection at F as

$$(20·84V - 171,600) \times 10^{-6} = 0, \quad \text{whence} \quad V = 8230\text{N}$$

Once V is known, the bar forces can be found from Fig. 69(d).

In each of the three examples discussed in this section so far, the truss had only one redundancy. The method of direct comparison of lengths can be applied to frames having any number of redundancies, but if the number exceeds one it will usually be found best to use the method of virtual work as a means of determining displacements, since calculations based on Williot–Mohr diagrams can become very cumbersome (for linear systems it is, however, possible to draw separate Williott–Mohr diagrams for the external load system and for each of the redundant forces and then obtain the solution by superposition). We conclude this section with an example of a truss having three redundancies.

The truss (a propped cantilever) is shown in Fig. 70(a). All bars have the same cross-sectional area and Young's modulus. Bar CG is made 0·3cm too long so that the truss is pre-stressed. We choose as the redundant forces those in bar BF (R_1kN), bar CG (R_2kN) and at support G (R_3kN). The forces in the bars of the statically determinate truss which remains when bars BF, CG and the support at G have been removed are shown in Fig. 70(b). The elongations due to these forces are given in the second column of the table of Fig. 70(c). The third column of the table shows the force (P') in each bar when a unit load is applied across BF and the fourth column

BRACED FRAMEWORKS

shows the product of this force and the elongation due to the load system of Fig. 70(b). In the remainder of the table the decrease in gap CG and the upward movement of joint G are similarly calculated. (Since the system is linear we have checks on our arithmetic—by

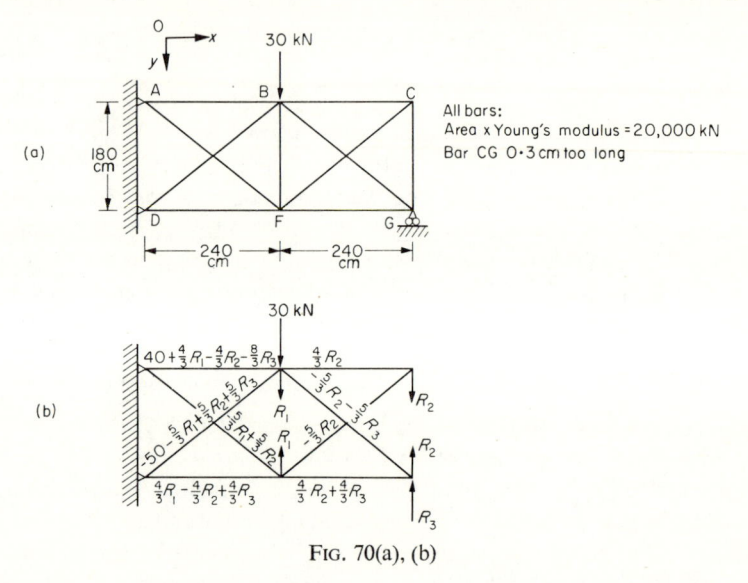

FIG. 70(a), (b)

Maxwell's reciprocal theorem the decrease in gap BF due to R_2 is equal to the decrease in gap CG due to R_1, and similarly for the two other pairs of reciprocal terms.)

Now, due to sustaining a force R_1, bar BF elongates by

$$\frac{R_1}{20,000} \times 180 = 9R_1 \times 10^{-3} \text{ cm}$$

We therefore have as our first compatibility equation,

$$\underbrace{180 - (1890 + 126R_1 - 126R_2 - 63R_3) \times 10^{-3}}_{\text{length of gap BF}} = \underbrace{180 + 9R_1 \times 10^{-3}}_{\text{length of bar BF}} \quad (33)$$

Due to force R_2, bar CG elongates by

$$\frac{R_2}{20,000} \times 180 = 9R_2 \times 10^{-3} \text{cm}$$

98

Bar	Elongation due to force of Fig. 70(b) (10^{-3} cm)	P' ($R_1=1$)	$P' \times$ elongation (10^{-3} cm)	P' ($R_2=1$)	$P' \times$ elongation (10^{-3} cm)	P' ($R_3=1$)	$P' \times$ elongation (10^{-3} cm)
AB	$480+16R_1+16R_2-32R_3$	$\frac{4}{3}$	$640+\frac{64}{3}R_1-\frac{64}{3}R_2-\frac{128}{3}R_3$	$-\frac{4}{3}$	$-640-\frac{64}{3}R_1+\frac{64}{3}R_2+\frac{128}{3}R_3$	$-\frac{8}{3}$	$-1280-\frac{128}{3}R_1+\frac{128}{3}R_2+\frac{256}{3}R_3$
BC	$16R_2$	0	0	$\frac{4}{3}$	$\frac{64}{3}R_2$	0	0
DF	$16R_1-16R_2+16R_3$	$\frac{4}{3}$	$\frac{64}{3}R_1-\frac{64}{3}R_2+\frac{64}{3}R_3$	$-\frac{4}{3}$	$-\frac{64}{3}R_1+\frac{64}{3}R_2-\frac{64}{3}R_3$	$\frac{4}{3}$	$\frac{64}{3}R_1-\frac{64}{3}R_2+\frac{64}{3}R_3$
FG	$16R_2+16R_3$	0	0	$\frac{4}{3}$	$\frac{64}{3}R_2+\frac{64}{3}R_3$	$\frac{4}{3}$	$\frac{64}{3}R_2+\frac{64}{3}R_3$
AF	$-25R_1+25R_2$	$-\frac{5}{3}$	$\frac{125}{3}R_1-\frac{125}{3}R_2$	$\frac{5}{3}$	$-\frac{125}{3}R_1+\frac{125}{3}R_2$	0	0
BD	$-750-25R_1+25R_2+25R_3$	$-\frac{5}{3}$	$1250+\frac{125}{3}R_1-\frac{125}{3}R_2-\frac{125}{3}R_3$	$\frac{5}{3}$	$-1250-\frac{125}{3}R_1+\frac{125}{3}R_2+\frac{125}{3}R_3$	$\frac{5}{3}$	$-1250-\frac{125}{3}R_1+\frac{125}{3}R_2+\frac{125}{3}R_3$
BG	$-25R_2-25R_3$	0	0	$-\frac{5}{3}$	$\frac{125}{3}R_2+\frac{125}{3}R_3$	$-\frac{5}{3}$	$\frac{125}{3}R_2+\frac{125}{3}R_3$
CF	$-25R_2$	0	0	$-\frac{5}{3}$	$\frac{125}{3}R_2$	0	0
		Total	$1890+126R_1-126R_2-63R_3$ =Decrease in gap BF	Total	$-1890-126R_1+252R_2+126R_3$ =Decrease in gap CG	Total	$-2530-63R_1+126R_2+\frac{634}{3}R_3$ =Upward movement of joint G

Fig. 70(c)

BRACED FRAMEWORKS

Bar CG was originally made 0·3cm too long. We have as our second compatibility equation,

$$\underbrace{180-(-1890-126R_1+252R_2+126R_3)\times 10^{-3}}_{\text{length of gap CG}} = \underbrace{180+9R_2\times 10^{-3}+0\cdot 3}_{\text{length of bar CG}} \quad (34)$$

Since support G does not move vertically, the upward movement of joint G must be zero, so the third compatibility equation is

$$\left(-2530-63R_1+126R_2+\frac{634}{3}R_3\right)\times 10^{-3} = 0 \quad (35)$$

Equations (33), (34) and (35) can be solved to yield $R_1 = -14\cdot 8$, $R_2 = -6\cdot 6$, $R_3 = 11\cdot 5$kN. The remaining bar forces can then be found from Fig. 70(b).

The method of solving redundant frameworks by the direct comparison of lengths, which has been described in this section, is of quite general application. The physical significance of each stage in the process of solution is clear, and errors of computation which lead to physical absurdities are easily discovered. The method does necessarily involve some direct consideration of geometrical compatibility, albeit of an elementary kind. The methods of solution described in the next section (energy methods) have been devised to avoid geometrical reasoning.

3.4. Energy Methods

3.4.1. *The energies of a structure.* In the present section we shall discuss a variety of methods for the analysis of frameworks. Some of these methods are useful in solving redundant frameworks; some can also be applied to statically determinate frames. Some of the methods are of general application; some can only be used in the analysis of structures which obey Hooke's law. All of the methods depend upon the concept of the energy of the structure. We shall begin by considering the different ways in which that energy may be defined.

Suppose we have a bar of datum length l_0 and that we apply to it a longitudinal force P. Let the force–elongation relationship be as shown in Fig. 71(a): the curve does not pass through the origin because the length at zero force may differ from l_0, due to initial

REDUNDANT TRUSSES

error or such causes as change of temperature or humidity. Suppose that the force in the bar is increased slowly from zero to P so that the bar remains at constant temperature during the process. Then two exchanges of energy take place during the increase of force. First, the forces P do work

$$u = \int_0^P P \cdot d(l-l_0) \qquad (36)$$

Second, there is a flow of heat Q into the bar from the surrounding atmosphere.

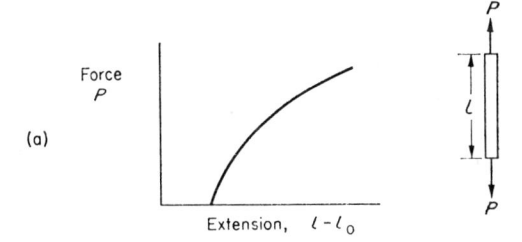

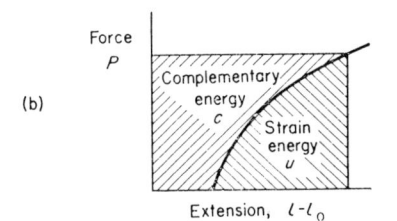

FIG. 71. Energies of a bar

u is known as the *strain energy* of the bar. It is equal to the area between the force–elongation curve and the axis of elongation (see Fig. 71(b)). Q can be determined by thermodynamic reasoning (for a linear elastic bar it is equal to $P\alpha Tl$, where T is the absolute temperature). Q is usually very much larger than u.

By considering conservation of energy, the *internal energy* of the bar increases by

$$\Delta u_i = u + Q \qquad (37)$$

BRACED FRAMEWORKS

One further definition of energy is required, that of *complementary energy*. This is defined as

$$c = \int_0^P (l-l_0)\,dP \tag{38}$$

It is equal to the area between the force–elongation curve and the axis of force (Fig. 71(b)).

u and c are of considerable value in structural problems. u_i is of importance in thermodynamics and when considering transient heating of high temperature structures, but we shall not refer to it again in the present book. u (the work done by the forces acting on the bar) is not generally recoverable in full when the bar is unloaded, unless the material is elastic.

On differentiating equations (36) and (38) we obtain

$$\frac{du}{d(l-l_0)} = P \tag{39}$$

and

$$\frac{dc}{dP} = l - l_0 \tag{40}$$

The concepts of energy can be extended from a single bar to a complete framework. By summation we obtain the strain energy of a framework as

$$U = \sum_{\text{bars}} \int_0^P P\,d(l-l_0) \tag{41}$$

and the complementary energy of a framework as

$$C = \sum_{\text{bars}} \int_0^P (l-l_0)\,dP \tag{42}$$

3.4.2. *Energy theorems*. The first group of energy theorems which are of use to us concern the relations between the energies of the structure, the external loads, and the displacements of the points of application of those loads. Suppose we have a framework in equilibrium under a set of *independent* loads $W_1 \ldots W_r \ldots W_n$, and that the displacements of the points of application of $W_1 \ldots W_r \ldots W_n$ in the directions of application of the loads are $y_1 \ldots y_r \ldots y_n$ (Fig. 72). It should be noted that $y_1 \ldots y_n$ are total displacements: they may depend partly on causes other than the application of $W_1 \ldots W_n$ (e.g. change of temperature of the bars, or initial errors of length). Now let $y_1 \ldots y_r \ldots y_n$ be increased by arbitrary small amounts $\delta y_1 \ldots \delta y_r \ldots \delta y_n$ and let there be corresponding small changes in the extensions

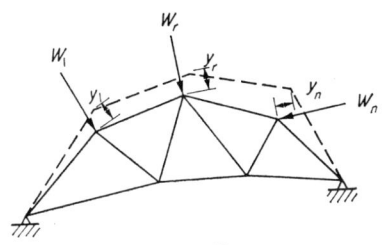

FIG. 72

of the bars, $\delta(l-l_0)$. Then the total work done on the joints of the frame during these small displacements is equal (see §2.8.3) to

$$W_1 \delta y_1 + \ldots W_r \delta y_r + \ldots W_n \delta y_n - \sum_{\text{bars}} P \cdot \delta(l-l_0)$$

and by the principle of virtual work this is zero. The term following the minus sign at the end of this expression is equal to the change in the strain energy of the frame, δU (see equation (41)). We therefore have

$$W_1 \delta y_1 + \ldots W_r \delta y_r + \ldots W_n \delta y_n - \delta U = 0 \quad (43)$$

Since $\delta y_1 \ldots \delta y_n$ are arbitrary, we can choose to make $\delta y_1 \ldots \delta y_{r-1}$ and $\delta y_{r+1} \ldots \delta y_n$ equal to zero. Equation (43) then becomes

$$W_r \cdot \delta y_r - \delta U = 0$$

or
$$W_r = \frac{\partial U}{\partial y_r} \quad (44)$$

If we can express the strain energy of a structure as a function of the displacements of the points of application of the external loads, then the partial differential coefficient of the strain-energy with respect to any one of those displacements is equal to the value of the corresponding load. Equation (44) was first derived by Castigliano (1879) for structures which obey Hooke's law: as will be seen by our present analysis, the equation is not in fact restricted to linear systems.†

The next relationship is obtained by the direct application of equation (24). A unit load applied at the same point and in the same direction as W_r will produce a force P' in a given bar equal to $\partial P/\partial W_r$, where P is the actual bar force. From equation (24) we therefore have

$$y_r = \sum_{\text{bars}} P'(l-l_0)$$

$$= \sum_{\text{bars}} \frac{\partial P}{\partial W_r}(l-l_0)$$

Substituting for $l-l_0$ from equation (40),

$$y_r = \sum_{\text{bars}} \frac{\partial P}{\partial W_r} \cdot \frac{dc}{dP}$$

$$= \frac{\partial C}{\partial W_r} \qquad (45)$$

If we can express the complementary energy of a structure as a function of the external loads, then the partial differential coefficient of the complementary energy with respect to any one of those loads is equal to the displacement of the loaded point in the direction of application of that load. It should be noted that equation (45) gives the total displacement, i.e. the displacement due to any initial changes in lengths of the bars together with the application of the load system. Equation (45) is due to Engesser (1889).

For a framework which is linear elastic so that in each of the bars the force–elongation curve is a straight line (Fig. 73), the strain energy and complementary energy are related by the equation

$$C = U + \sum_{\text{bars}} P \cdot \lambda \qquad (46)$$

where $l_0 + \lambda$ is the length of a bar at zero force.

---

† Equation (44) is equivalent to the theorem of minimum total potential, since if V is the potential energy of the loads we have

$$\text{Total potential} = \varphi = U + V$$

$$\frac{\partial \varphi}{\partial y_r} = \frac{\partial U}{\partial y_r} + \frac{\partial V}{\partial y_r} = W_r - W_r = 0$$

From equation (45) we have

$$y_r = \frac{\partial C}{\partial W_r} = \frac{\partial U}{\partial W_r} + \sum_{\text{bars}} \frac{\partial P}{\partial W_r} \cdot \lambda$$

$$= \frac{\partial U}{\partial W_r} + \sum P' \cdot \lambda \qquad (47)$$

The final term in this equation represents the part of y_r which is due to the initial changes (λ) in the lengths of the members. The term $\partial U/\partial W_r$ represents the part of y_r which is due to the application of the load system. If we denote this by $\Delta_{r(W_1 \cdots W_n)}$, we have

$$\Delta_{r(W_1 \cdots W_n)} = \frac{\partial U}{\partial W_r} \qquad (48)$$

If we can express the strain energy of a structure which obeys Hooke's law as a function of the external loads, then the partial differential coefficient of the strain energy with respect to any one of those loads

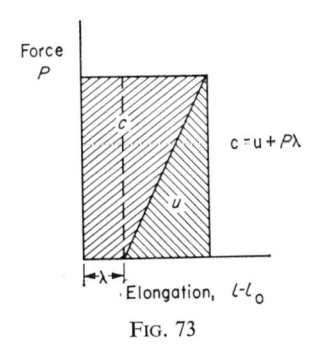

FIG. 73

is equal to that part of the displacement of the loaded joint, in the direction of application of the load, which is due to the application of the load system.

Equation (48) was first stated by Castigliano (1879). It is known under a variety of names, some authors referring to it as "the first theorem of Castigliano" and others as "Castigliano's theorem, Part II" (Part I being equation (44)). It is probably the most used and misused theorem in the whole of structural analysis. This theorem of Castigliano is only true for structures which obey Hooke's law. The deflection which it gives is not the total deflection of the point

BRACED FRAMEWORKS

concerned, but only that part of it which is due to the application of the applied load system (in many problems, of course, this will be the whole deflection). This theorem of Castigliano has advantages in certain types of problem over Engesser's equation (45), but its use is very restricted whereas equation (45) is of general validity. The reader is therefore advised to avoid employing Castigliano's theorem and whenever possible to use equation (45) instead.

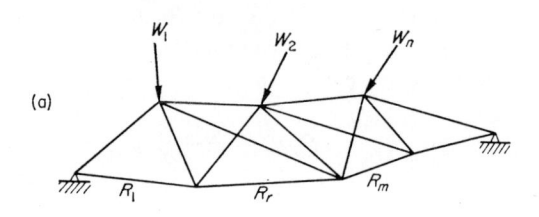

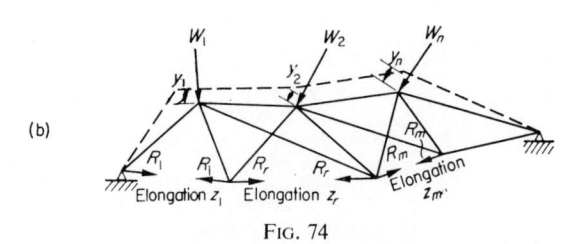

Fig. 74

The second group of energy theorems are concerned with the relations between the energies of a redundant structure and the values of the redundant forces. Suppose we have a framework subjected to external loads $W_1 \ldots W_n$ and internal redundant constraints $R_1 \ldots R_r \ldots R_m$ (Fig. 74(a)). Let the displacements of the points of application of the loads in the directions of load application be $y_1 \ldots y_n$ and let the elongations of the members supporting $R_1 \ldots R_r \ldots R_m$ be $z_1 \ldots z_r \ldots z_m$. Remove the redundant members and consider the statically determinate frame which remains (Fig. 74(b)). Let its strain energy U' be expressed in terms of $y_1 \ldots y_n$ and $z_1 \ldots z_m$. Then from equation (44),

$$R_r = -\frac{\partial U'}{\partial z_r} \tag{49}$$

(the minus sign occurs because R and z are measured in opposite directions).

Now consider bar r. If we denote its strain energy by u_r, then, from equation (39),

$$\frac{du_r}{dz_r} = R_r \qquad (50)$$

If we denote the strain energy of any other redundant bar t by u_t, u_t is a function of z_t only. It follows that

$$\frac{\partial u_t}{\partial z_r} = 0 \quad (t \neq r) \qquad (51)$$

The strain energy of the whole frame, including the redundant members, is given by

$$U = U' + u_r + \sum_{t=1}^{t=r-1} u_t + \sum_{t=r+1}^{t=m} u_t$$

Using equations (49), (50) and (51),

$$\frac{\partial U}{\partial z_r} = \frac{\partial U'}{\partial z_r} + \frac{du_r}{dz_r} + \sum_{t=1}^{t=r-1} \frac{\partial u_t}{\partial z_r} + \sum_{t=r+1}^{t=m} \frac{\partial u_t}{\partial z_r}$$
$$= -R_r + R_r$$
$$= 0 \qquad (52)$$

If we can express the strain energy of a redundant structure as a function of the displacements of the points of application of the external loads and the elongations of the redundant members, then the partial differential coefficient of the strain energy with respect to the elongation of any one of the redundant members is zero.

A similar theorem can be developed using complementary energy. From equation (45),

$$z_r = -\frac{dC'}{dR_r} \qquad (53)$$

where C' is the complementary energy of the statically determinate frame which remains when the redundant members have been removed. C' is expressed in terms of $W_1 \ldots W_n$ and $R_1 \ldots R_m$.

For bar r we have from equation (40),

$$\frac{dc_r}{dR_r} = z_r \qquad (54)$$

BRACED FRAMEWORKS

Since the complementary energy of any other redundant bar t is a function of R_t only,

$$\frac{\partial c_t}{\partial R_r} = 0 \quad (t \neq r) \tag{55}$$

The complementary energy of the whole frame

$$C = C' + c_r + \sum_{t=1}^{t=r-1} c_t + \sum_{t=r+1}^{t=m} c_t$$

Using equations (53), (54) and (55),

$$\begin{aligned}\frac{\partial C}{\partial R_r} &= \frac{\partial C'}{\partial R_r} + \frac{dc_r}{dR_r} + \sum_{t=1}^{t=r-1} \frac{\partial c_t}{\partial R_r} + \sum_{t=r+1}^{t=m} \frac{\partial c_t}{\partial R_r} \\ &= -z_r + z_r \\ &= 0 \end{aligned} \tag{56}$$

If we can express the complementary energy of a redundant structure as a function of the external loads and the forces in the redundant members, then the partial differential coefficient of the complementary energy with respect to the force in any one of the redundant members is zero.

For a framework which obeys Hooke's law we may use equation (46) to obtain

$$\frac{\partial U}{\partial R_r} = \frac{\partial C}{\partial R_r} - \sum_{\text{bars}} \frac{\partial P}{\partial R_r} \cdot \lambda \tag{57}$$

If we define the datum geometry as that of the unloaded statically determinate frame, λ is zero except in the redundant bars. In redundant bar t, $P = R_t$. Since R_r and R_t are independent variables, it follows that $\partial P/\partial R_r = 0$ ($t \neq r$). In bar r, $P = R_r$ and $\partial P/\partial R_r = 1$. Using these relationships and equation (56), equation (57) becomes

$$\frac{\partial U}{\partial R_r} = -\lambda_r \tag{58}$$

If we can express the strain energy of a structure which obeys Hooke's law as a function of the external loads and the forces in the redundant members, then the partial differential coefficient of the strain energy with respect to the force in any one of the redundant members is equal to minus the initial excess of length of that member. (It is understood that the datum geometry is taken as that of the unloaded statically determinate structure which remains when the redundant members are removed.) Equation (58), with $\lambda_r = 0$, was first stated without proof by Ménabréa (1858). The proof was supplied later by Castigliano (1879), who derived it from equation (48). It is called by some authors "the second theorem of Castigliano". Equation (58) has slight computational advantages over equation (56) in certain types of problem, but like equation (48) it is of limited validity. The reader is therefore again advised to eschew Castigliano's theorem and to use the general form of equation (56) instead.

In most problems of structural analysis it is much easier to express the energy of the structure in terms of forces, rather than in terms of displacements. Of the various energy theorems derived above, the two most useful are therefore those of equations (45) and (56). We shall refer to these as

The first complementary energy theorem,

$$\frac{\partial C}{\partial W_r} = y_r \qquad (45)$$

and

The second complementary energy theorem,

$$\frac{\partial C}{\partial R_r} = 0 \qquad (56)$$

For structures obeying Hooke's law, these may be replaced (without much advantage) by

The first strain energy theorem (also known as Castigliano's theorem Part II, or the first theorem of Castigliano)
$$\frac{\partial U}{\partial W_r} = \Delta_r(w_1 \ldots w_n) \qquad (48)$$

and

The second strain energy theorem, (also known as the second theorem of Castigliano)
$$\frac{\partial U}{\partial R_r} = -\lambda_r \qquad (58)$$

(with limitations on the datum for measurement—see above).

The advantage claimed by their adherents for Castigliano's theorems, which make use of strain energy, is that only the squares of bar forces are involved, so that mistakes in signs are unimportant.

BRACED FRAMEWORKS

This argument is of doubtful validity, and for the case of members having no initial changes of length the complementary energy has the same property. For frames in which the members have initial changes of length, due to rise of temperature, etc., the second strain energy theorem can become extremely tedious to use because of the difficulty of defining the datum geometry: the second complementary energy theorem does not suffer from this limitation.

3.4.3. *The energies of linear-elastic bars.* Consider a bar of datum length l_0, cross-sectional area A and Young's modulus E subjected to a longitudinal force P. The stress in the bar is P/A, the strain P/AE and the elongation $\{(Pl_0/AE)+\lambda\}$ where λ is the elongation at zero force (due to temperature rise, errors of manufacture, etc.).

Then the strain energy of the bar is given by equation (36) as

$$u = \int_0^P P \, . \, d\left(\frac{Pl_0}{AE}+\lambda\right) = \frac{P^2 l_0}{2AE} \qquad (59)$$

and the complementary energy of the bar is given by equation (38) as

$$c = \int_0^P \left(\frac{Pl_0}{AE}+\lambda\right) dP = \frac{P^2 l_0}{2AE}+P\lambda \qquad (60)$$

3.5. Solution by Energy Methods

In the present section we show applications of the energy theorems discussed in §3.4.2. The reader should always remember that the purpose of energy methods is to provide means of solving geometrical problems without the direct consideration of geometry. In any case of doubt as to the correctness or validity of an energy solution, the difficulty can always be resolved by reverting to the fundamental geometrical reasoning of §3.3.

We begin by considering the truss shown in Fig. 75(a), which is statically determinate. It is desired to find the vertical deflection at G due to the load system shown, together with initial increases in length of members BK and DH of 0·05cm. If we denote the load at G by W, the forces in the bars are as shown in Fig. 75(b). These forces are tabulated in Fig. 75(c), together with the lengths of the members and the initial extensions. The final column of the table

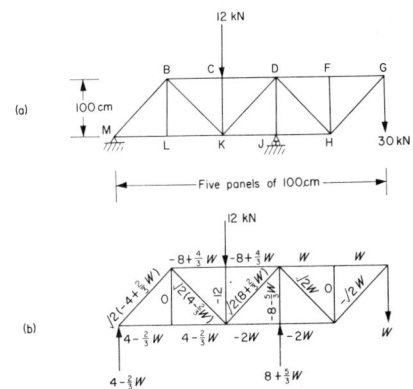

(a) 12 kN applied at C; truss with joints B, C, D, F, G on top and M, L, K, J, H on bottom; 100 cm height; 30 kN at H; Five panels of 100 cm.

All bars: area × Young's modulus = 80,000 kN
BK and DH are initially 0·05 cm too long

(b) Force diagram with member forces:
- Top chord: $-8+\frac{4}{3}W$, $-8+\frac{4}{3}W$, W, W
- Diagonals: $\sqrt{2}(-4+\frac{2}{3}W)$, 0, $-\sqrt{2}(4-\frac{2}{3}W)$, $-\frac{1}{\sqrt{2}}(8+\frac{4}{3}W)$, $\sqrt{2}W$, 0, $-\sqrt{2}W$
- Bottom chord: $4-\frac{2}{3}W$, $4-\frac{2}{3}W$, $-2W$, $-2W$
- Reactions: $4-\frac{2}{3}W$ (left), $8+\frac{5}{3}W$ (right), W

Member	Force (P) (kN)	Length (l_0) (cm)	Initial extn. λ (cm)	Complementary energy $(P^2 l_0/2AE) + P\lambda$ (kN. cm)
MB	$\sqrt{2}\left(-4+\frac{2}{3}W\right)$	$100\sqrt{2}$	0	$\left(2\sqrt{2}-\frac{2}{3}\sqrt{2}\,W+\frac{\sqrt{2}}{18}W^2\right)\times 10^{-2}$
BC	$-8+\frac{4}{3}W$	100	0	$\left(4-\frac{4}{3}W+\frac{1}{9}W^2\right)\times 10^{-2}$
CD	$-8+\frac{4}{3}W$	100	0	$\left(4-\frac{4}{3}W+\frac{1}{9}W^2\right)\times 10^{-2}$
DF	W	100	0	$\frac{1}{16}W^2 \times 10^{-2}$
FG	W	100	0	$\frac{1}{16}W^2 \times 10^{-2}$
ML	$4-\frac{2}{3}W$	100	0	$\left(1-\frac{1}{3}W+\frac{1}{36}W^2\right)\times 10^{-2}$
LK	$4-\frac{2}{3}W$	100	0	$\left(1-\frac{1}{3}W+\frac{1}{36}W^2\right)\times 10^{-2}$
KJ	$-2W$	100	0	$\frac{1}{4}W^2 \times 10^{-2}$
JH	$-2W$	100	0	$\frac{1}{4}W^2 \times 10^{-2}$
HG	$-\sqrt{2}W$	$100\sqrt{2}$	0	$\frac{\sqrt{2}}{8}W^2 \times 10^{-2}$
BL	0	100	0	0
CK	-12	100	0	9×10^{-2}
DJ	$-8-\frac{5}{3}W$	100	0	$\left(4+\frac{5}{3}W+\frac{25}{144}W^2\right)\times 10^{-2}$
FH	0	100	0	0
BK	$\sqrt{2}\left(4-\frac{2}{3}W\right)$	$100\sqrt{2}$	0·05	$\left(2\sqrt{2}-4\sqrt{2}\,W+\frac{\sqrt{2}}{18}W^2\right)\times 10^{-2}$
DK	$\sqrt{2}\left(8+\frac{2}{3}W\right)$	$100\sqrt{2}$	0	$\left(8\sqrt{2}+\frac{4}{3}\sqrt{2}\,W+\frac{\sqrt{2}}{18}W^2\right)\times 10^{-2}$
DH	$\sqrt{2}W$	$100\sqrt{2}$	0·05	$\left(5\sqrt{2}\,W+\frac{\sqrt{2}}{8}W^2\right)\times 10^{-2}$
GH	$-\sqrt{2}W$	$100\sqrt{2}$	0	$\frac{\sqrt{2}}{8}W^2 \times 10^{-2}$

(c)

Total $\{23 + 32\sqrt{2} + \frac{5}{3}(\sqrt{2}-1)W + \frac{1}{144}(155 + 70\sqrt{2})W^2\} \times 10^{-2}$

Fig. 75

111

BRACED FRAMEWORKS

shows the complementary energy of each bar, calculated from equation (60).

The complementary energy of the whole truss,

$$C = \{23 + 32\sqrt{2} + \tfrac{5}{3}(\sqrt{2}-1)W + \\ + \tfrac{1}{144}(155 + 70\sqrt{2})W^2\} \times 10^{-2} \text{ kN . cm}$$

From the first complementary energy theorem (equation (45)), the deflection at **G**

$$y = \frac{\partial C}{\partial W} = \{\tfrac{5}{3}(\sqrt{2}-1) + \tfrac{1}{72}(155 + 70\sqrt{2})W\} \times 10^{-2} \text{ cm}$$

Putting W equal to 30kN, we find $y = 1.07$cm.

Since we are eventually concerned with $\partial C/\partial W$, rather than with C, the calculation could be slightly simplified by determining $\partial C/\partial W$ for the individual bars. From equation (40)

$$\frac{\partial c}{\partial W_r} = \frac{dc}{dP} \cdot \frac{\partial P}{\partial W_r} = (l - l_0) \cdot \frac{\partial P}{\partial W_r} \qquad (61)$$

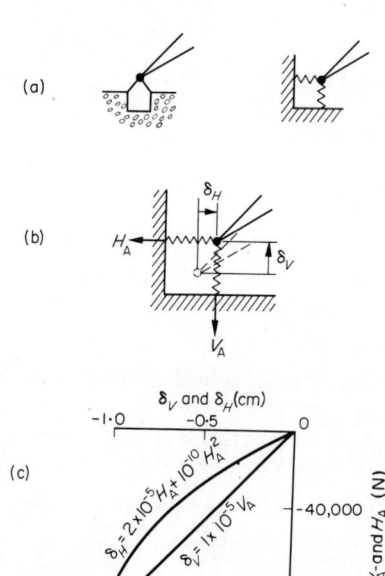

FIG. 76. Flexible supports

As $l-l_0$ is equal to the extension of the bar, and $\partial P/\partial W_r$ is the bar force due to a unit load W_r, the method is clearly equivalent to the determination of deflections by virtual work discussed in §2.8.3.

As a second example, consider the arch shown in Fig. 77(a). The stress–strain relationship for the material of the bars is given with sufficient accuracy by the equation

$$\varepsilon = 10^{-7}\sigma(1+10^{-9}\sigma^2) \tag{62}$$

where σ is in N/cm². Support F is fixed, but support A deflects under vertical and horizontal load according to the equations

$$\left.\begin{array}{l}\delta_V = 1 \times 10^{-5} V_A \\ \delta_H = 2 \times 10^{-5} H_A + 10^{-10} H_A^2\end{array}\right\} \tag{63}$$

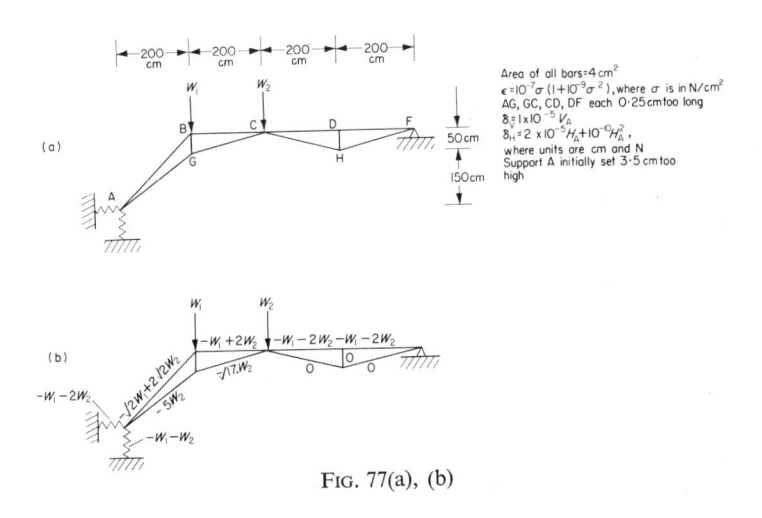

FIG. 77(a), (b)

where δ_V and δ_H are in cm and V_A and H_A are in N. The most convenient way of treating a flexible support analytically is to regard it as consisting of two very short members connecting the frame to fixed points, as shown in Fig. 76(a). Equations (63) then refer to the force–elongation characteristics of these members. Since we conventionally make tensile force and elongation positive, the signs in equation (63) must be chosen accordingly (Fig. 76(b)), even though the experimental data on which the equations are based have been obtained from tests in compression (Fig. 76(c)).

Members AG, GC, CD and DF are each made 0·25cm too long.

BRACED FRAMEWORKS

Member	Stress σ (N/cm²)	Elongation due to stress (cm)
AB	$-\dfrac{1}{2\sqrt{2}}W_1 + \dfrac{1}{\sqrt{2}}W_2$	$10^{-5}(-W_1 + 2W_2) + $ $+ 10^{-14}\left(-\dfrac{1}{8}W_1^3 + \dfrac{3}{4}W_1^2 W_2 - \dfrac{3}{2}W_1 W_2^2 + W_2^3\right)$
BC	$-\dfrac{1}{4}W_1 + \dfrac{1}{2}W_2$	$10^{-5}\left(-\dfrac{1}{2}W_1 + W_2\right) + $ $+ 10^{-14}\left(-\dfrac{1}{32}W_1^3 + \dfrac{3}{16}W_1^2 W_2 - \dfrac{3}{8}W_1 W_2^2 + \dfrac{1}{4}W_2^3\right)$
CD	$-\dfrac{1}{4}W_1 - \dfrac{1}{2}W_2$	$10^{-5}\left(-\dfrac{1}{2}W_1 - W_2\right) + $ $+ 10^{-14}\left(-\dfrac{1}{32}W_1^3 - \dfrac{3}{16}W_1^2 W_2 - \dfrac{3}{8}W_1 W_2^2 - \dfrac{1}{4}W_2^3\right)$
DF	$-\dfrac{1}{4}W_1 - \dfrac{1}{2}W_2$	$10^{-5}\left(-\dfrac{1}{2}W_1 - W_2\right) + $ $+ 10^{-14}\left(-\dfrac{1}{32}W_1^3 - \dfrac{3}{16}W_1^2 W_2 - \dfrac{3}{8}W_1 W_2^2 - \dfrac{1}{4}W_2^3\right)$
AG	$-\dfrac{5}{4}W_2$	$10^{-5}\left(-\dfrac{25}{8}W_2\right) + $ $+ 10^{-14} \qquad\qquad\qquad \left(-\dfrac{625}{128}W_2^3\right)$
GC	$-\dfrac{\sqrt{17}}{4}W_2$	$10^{-5}\left(-\dfrac{17}{8}W_2\right) + $ $+ 10^{-14} \qquad\qquad\qquad \left(-\dfrac{289}{128}W_2^3\right)$
CH	0	0
HF	0	0
BG	$-\dfrac{1}{2}W_2$	$10^{-5}\left(-\dfrac{1}{4}W_2\right) + $ $+ 10^{-14} \qquad\qquad\qquad \left(-\dfrac{1}{16}W_2^3\right)$
DH	0	

Support A		Force (N)	Elongation due to force (cm)
	Vert.	$-W_1 - W_2$	$10^{-5}(-W_1 - W_2)$
	Hor.	$-W_1 - 2W_2$	$10^{-5}(-2W_1 - 4W_2) + 10^{-10}(W_1^2 + 4W_1 W_2 + 4W_2^2)$

Fig. 77(c

Elongation due to other causes (cm)	$\dfrac{\partial P}{\partial W_2}$	Total elongation $\times \dfrac{\partial P}{\partial W_2} = \dfrac{\partial c}{\partial W_2}$ (cm)
0	$2\sqrt{2}$	$10^{-5}(-2\sqrt{2}W_1 + 4\sqrt{2}W_2) +$ $+ 10^{-14}\left(-\dfrac{\sqrt{2}}{4}W_1^3 + \dfrac{3\sqrt{2}}{2}W_1^2 W_2 - 3\sqrt{2}W_1 W_2^2 + 2\sqrt{2}W_2^3\right)$
0	2	$10^{-5}(-W_1 + 2W_2) +$ $+ 10^{-14}\left(-\dfrac{1}{16}W_1^3 + \dfrac{3}{8}W_1^2 W_2 - \dfrac{3}{4}W_1 W_2^2 + \dfrac{1}{2}W_2^3\right)$
0·25	-2	$-0{\cdot}5 + 10^{-5}(W_1 + 2W_2) +$ $+ 10^{-14}\left(\dfrac{1}{16}W_1^3 + \dfrac{3}{8}W_1^2 W_2 + \dfrac{3}{4}W_1 W_2^2 + \dfrac{1}{2}W_2^3\right)$
0·25	-2	$-0{\cdot}5 + 10^{-5}(W_1 + 2W_2) +$ $+ 10^{-14}\left(\dfrac{1}{16}W_1^3 + \dfrac{3}{8}W_1^2 W_2 + \dfrac{3}{4}W_1 W_2^2 + \dfrac{1}{2}W_2^3\right)$
0·25	-5	$-1{\cdot}25 + 10^{-5}\left(\dfrac{125}{8}W_2\right) +$ $+ 10^{-14} \qquad \left(\dfrac{3125}{128}W_2^3\right)$
0·25	$-\sqrt{17}$	$-0{\cdot}25\sqrt{17} + 10^{-5}\left(\dfrac{17\sqrt{17}}{8}W_2\right) +$ $+ 10^{-14} \qquad \left(\dfrac{289\sqrt{17}}{128}W_2^3\right)$
0	0	0
0	0	0
0	-2	$10^{-5}\left(\dfrac{1}{2}W_2\right) +$ $+ 10^{-14} \qquad \left(\dfrac{1}{8}W_2^3\right)$
0	0	0
3·5	-1	$-3{\cdot}5 + 10^{-5}(W_1 + W_2)$
0	-2	$10^{-5}(4W_1 + 8W_2) + 10^{-10}(-2W_1^2 - 8W_1 W_2 - 8W_2^2)$
	Total	$-6{\cdot}78 + 10^{-5}(3{\cdot}17 W_1 + 45{\cdot}55 W_2) + 10^{-10}(-2W_1^2 - 8W_1 W_2 - 8W_2^2)$ $+ 10^{-14}(-0{\cdot}291 W_1^3 + 3{\cdot}246 W_1^2 W_2 - 3{\cdot}492 W_1 W_2^2 + 38{\cdot}19 W_2^3)$

BRACED FRAMEWORKS

Support A is set initially 3·5cm too high: this is equivalent to the short vertical bar being initially 3·5cm too long.

It is desired to find the vertical deflection of C for the range of values W_1 (0–40,000N) and W_2 (0–20,000N). The forces in the bars are shown in Fig. 77(b), and the stresses in the second column of the table of Fig. 77(c). The elongations of the bars, calculated by multiplying the strains (from equation (62)) by the lengths, or in the case of the flexible support directly from equations (63), are shown

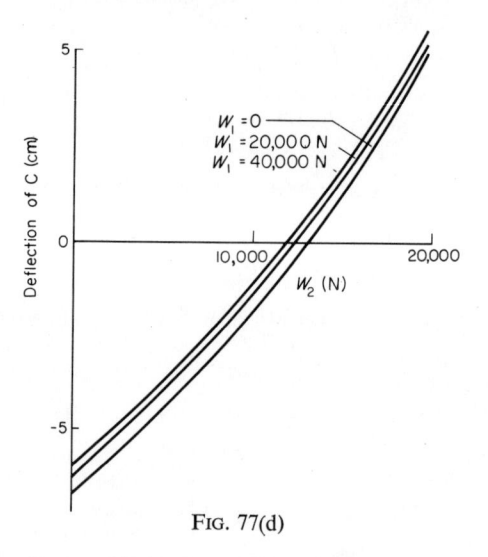

Fig. 77(d)

in the third column of the table. The fourth column gives the initial elongations of the bars, and in the fifth column the value of the bar force per unit load at C ($= \partial P/\partial W_2$) is calculated from Fig. 77(b). The final column of the table gives the total elongation multiplied by $\partial P/\partial W_2$: from equation (61) this is equal to $\partial c/\partial W_2$.

The total of $\partial c/\partial W_2$ for all bars of the frame (including those representing the flexible support) is equal to the deflection at C (from the first complementary energy theorem). We find

$$y = \frac{\partial c}{\partial W_2} = -6\cdot78 + (3\cdot17W_1 + 45\cdot55W_2) \times 10^{-5} +$$
$$+ (-2W_1^2 - 8W_1 W_2 - 8W_2^2) \times 10^{-10} + (-0\cdot291 W_1^3 +$$
$$+ 3\cdot246 W_1^2 W_2 - 3\cdot492 W_1 W_2^2 + 38\cdot19 W_2^3) \times 10^{-14}.$$

The values of the deflection are plotted in Fig. 77(d).

REDUNDANT TRUSSES

As our third example we consider the redundant truss shown in Fig. 78(a). The truss is initially free from stress and it is then subjected to the given load and temperature rises. The truss has two redundancies. If we choose these as the forces in FD and DC, the bar forces are as shown in Fig. 78(b). These bar forces are tabulated in Fig. 78(c), together with the cross-sectional areas and lengths of the members. The fifth column of the table shows λ, the change in length of the member from the datum value due to causes other than the application of force: in the present example $\lambda = \alpha T l_0$, where T is the temperature rise.

In the final column of the table of Fig. 78(c), the complementary energy of each bar is calculated from equation (60). The complementary energy of the whole truss,

$$C = 0{\cdot}641 + 0{\cdot}0285 R_1^2 + 0{\cdot}0856 R_2^2 + 0{\cdot}1902 R_1 -$$
$$- 0{\cdot}4541 R_2 - 0{\cdot}0820 R_1 R_2 \text{ kN . cm} \quad (64)$$

Now from the second complementary energy theorem (equation (56)) $\partial C/\partial R_1 = \partial C/\partial R_2 = 0$. Substituting the value of C from equation (64), we obtain two simultaneous equations in R_1 and R_2 which can be solved to give $R_1 = 1{\cdot}54\text{kN}$, $R_2 = 3{\cdot}39\text{kN}$. The remaining bar forces can now be found from Fig. 78(b).

For three-dimensional frames it will often be convenient to express the complementary energy in terms of tension coefficients. From equations (13) and (60),

$$c = \frac{t^2 l_0^3}{2AE} + \lambda l_0 t \quad (65)$$

Consider the frame shown in Fig. 79(a), which has one redundancy. Bars AF, FH, and CJ are initially 0·01cm too short and the frame is subjected to a vertical load of 200N at F. If we choose bar FH as the redundant member and denote its tension coefficient by t, the tension coefficients of all the bars can readily be found by using equations (15). The tension coefficients are shown in Fig. 79(a) and in the second column of the table of Fig. 79(b). The third column of the table gives the length of each bar, and the fourth column the initial error in length. In the fifth column the complementary energy is calculated from equation (65).

117

BRACED FRAMEWORKS

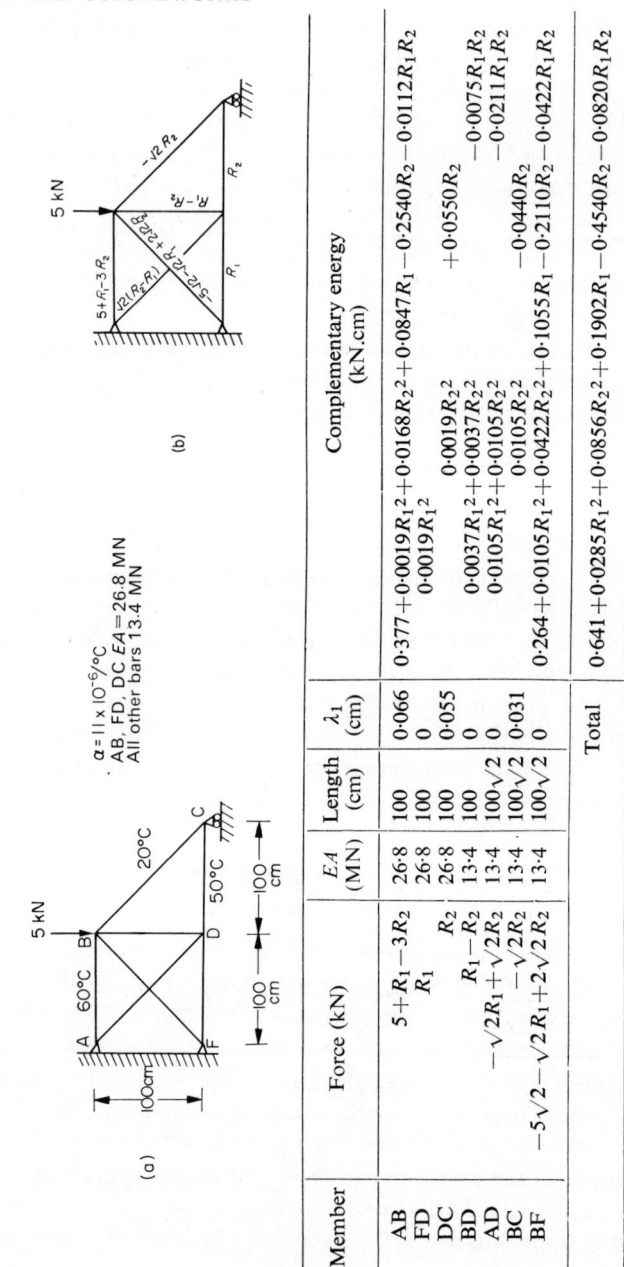

$\alpha = 11 \times 10^{-6}/°C$
AB, FD, DC $EA = 26.8$ MN
All other bars 13.4 MN

Member	Force (kN)	EA (MN)	Length (cm)	λ_1 (cm)	Complementary energy (kN.cm)
AB	$5+R_1-3R_2$	26.8	100	0.066	$0.377+0.0019R_1^2+0.0168R_2^2+0.0847R_1-0.2540R_2-0.0112R_1R_2$
FD	R_1	26.8	100	0	$0.0019R_1^2$
DC	R_2	26.8	100	0.055	$0.0019R_2^2 \qquad +0.0550R_2$
BD	R_1-R_2	13.4	100	0	$0.0037R_1^2+0.0037R_2^2 \qquad -0.0075R_1R_2$
AD	$-\sqrt{2}R_1+\sqrt{2}R_2$	13.4	$100\sqrt{2}$	0	$0.0105R_1^2+0.0105R_2^2 \qquad -0.0211R_1R_2$
BC	$-\sqrt{2}R_2$	13.4	$100\sqrt{2}$	0.031	$0.0105R_2^2 \qquad -0.0440R_2$
BF	$-5\sqrt{2}-\sqrt{2}R_1+2\sqrt{2}R_2$	13.4	$100\sqrt{2}$	0	$0.264+0.0105R_1^2+0.0422R_2^2+0.1055R_1-0.2110R_2-0.0422R_1R_2$
Total					$0.641+0.0285R_1^2+0.0856R_2^2+0.1902R_1-0.4540R_2-0.0820R_1R_2$

Fig. 78

REDUNDANT TRUSSES

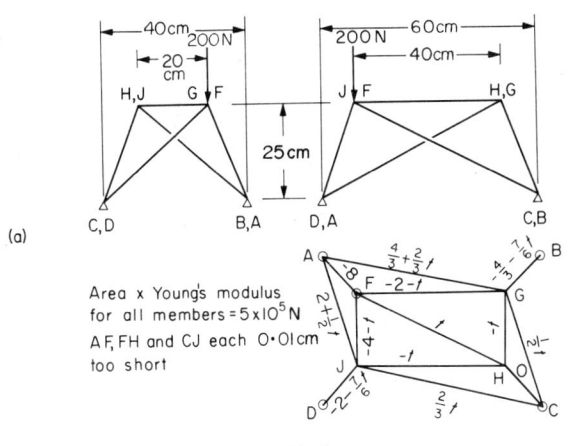

(a)

Fig. 79(a)

The complementary energy of the whole frame,

$$C = 4 \cdot 93 + 0 \cdot 135t + 0 \cdot 492t^2 \text{ N.cm}$$

By the second complementary energy theorem, $\partial C/\partial t = 0$, whence $t = -0 \cdot 137 \text{N/cm}$. Substituting this value of t in the second column of the table of Fig. 79(b), we obtain the tension coefficients given in the sixth column. The forces in the bars are obtained by multiplying the tension coefficients by the bar lengths.

In the two previous examples on redundant frames, we have calculated the complete complementary energy. Since we are eventually concerned with differential coefficients, it will often be simpler to determine these directly. From equation (40),

$$\frac{\partial c}{\partial R_r} = \frac{dc}{dP} \cdot \frac{\partial P}{\partial R_r} = (l - l_0) \cdot \frac{\partial P}{\partial R_r} \qquad (66)$$

As in the case of equation (61), the method is now equivalent to that of virtual work.

It will be seen from the examples in this section and in §3.3 that energy methods or direct geometrical methods for the analysis of redundant structures lead to the solution of a group of m simultaneous equations (in general non-linear) of the form $\partial C/\partial R_1 = 0$, $\partial C/\partial R_2 = 0, \ldots \partial C/\partial R_m = 0$, where $R_1 \ldots R_m$ are the unknown

BRACED FRAMEWORKS

Member	Tension coefficient (N/cm)	Length (cm)	λ (cm)	Complementary energy (N.cm)	Tension coefficient (N/cm)	Force (N)
AF	-8	$5\sqrt{33}$	-0.01	3.82	-8	-230
BG	$-\frac{4}{3}-\frac{7}{6}t$	$5\sqrt{33}$	0	$0.04+0.074t+0.032t^2$	-1.173	-34
CH	0	$5\sqrt{33}$	0	0	0	0
DJ	$-2-\frac{7}{6}t$	$5\sqrt{33}$	0	$0.09+0.111t+0.032t^2$	-1.840	-53
AG	$\frac{4}{3}+\frac{2}{3}t$	$5\sqrt{129}$	0	$0.33+0.326t+0.081t^2$	1.242	71
CJ	$\frac{2}{3}t$	$5\sqrt{129}$	-0.01	$-0.379t+0.081t^2$	-0.091	-5
AJ	$2+\frac{1}{3}t$	$5\sqrt{65}$	0	$0.26+0.131t+0.016t^2$	1.931	78
CG	$\frac{1}{3}t$	$5\sqrt{65}$	0	$0.016t^2$	-0.069	-3
FG	$-2-t$	40	0	$0.26+0.256t+0.064t^2$	-1.863	-75
HJ	$-t$	40	0	$0.064t^2$	0.137	5
GH	$-t$	20	0	$0.008t^2$	0.137	5
JF	$-4-t$	20	0	$0.13+0.064t+0.008t^2$	-3.863	-77
FH	t	$10\sqrt{20}$	-0.01	$-0.448t+0.090t^2$	-0.137	-6
			Total	$4.93+0.135t+0.492t^2$		

Fig. 79(b)

redundancies. The labour of solving these m equations can be very laborious (for $m > 6$ some form of automatic computer is almost essential). In the next section we consider methods devised to avoid simultaneous equations.

3.6. Relaxation Methods

The methods of structural analysis to be described in the present section avoid both the direct consideration of geometry and the solution of simultaneous equations. Unlike energy methods, however, relaxation methods enable the solver to visualise the physical behaviour of the structure very readily. The methods are essentially arithmetic: they are not suitable for algebraic solution.

Consider a pin-jointed redundant framework in which all of the pins are held rigidly in their datum positions by adjustable screw-jacks. If the bars are all of datum length at zero force, the bars will be free from stress and they will impose no load on the jacks. If the bars have initial errors of length, they will be stressed and will apply loads to the jacks which can readily be calculated. Now let external loads be applied to the joints of the framework. Since the joints are prevented from moving, these loads are sustained entirely by the screw-jacks, and there is no change in the forces (if any) in the bars.

Now consider that screw-jack which sustains the largest load. Let the jack be moved in the direction of the load. Then the load applied to the jack will decrease (to zero if we choose the correct displacement). Due to the movement of the joint, the forces in the members connected to it will change and the loads applied to the jacks at the far ends of these members will also change.

We choose the jack which now sustains the largest load and repeat the procedure. This process is continued until the loads sustained by the jacks are negligible. The external loads have now been transferred to the framework and the solution of the problem is complete. This analysis by the successive relaxation of constraints (Southwell, 1940) requires for its application a knowledge of the components of load applied by a bar to the jacks at its ends when one end of the bar is subjected to given displacements. We shall term this "the unit problem".

Suppose we have a bar AB, of original length l_{AB}. Let the positions

BRACED FRAMEWORKS

of A and B be defined by Cartesian coordinates and let the projections of l_{AB} on the coordinate axes be x_{AB}, y_{AB}, and z_{AB}. Then if joint A suffers displacements u_A, v_A and w_A in the directions of the coordinate axes, the extension of bar AB is given by equation (23) as

$$\delta l_{AB} = -\frac{1}{l_{AB}}(x_{AB} u_A + y_{AB} v_A + z_{AB} w_A) \qquad (67)$$

If bar AB is elastic, having Young's modulus E_{AB} and uniform cross-sectional area A_{AB}, the force produced in it by the displacements of joint A will be

$$P_{AB} = E_{AB} A_{AB} \frac{\delta l_{AB}}{l_{AB}}$$

$$= \frac{-E_{AB} A_{AB}}{l_{AB}^2}(x_{AB} u_A + y_{AB} v_A + z_{AB} w_A) \qquad (68)$$

The components of load applied to the jacks at joint A by bar AB can be obtained by resolving P_{AB} in the directions of the coordinate axes as

$$\left.\begin{aligned}
X_{A(AB)} &= \frac{x_{AB}}{l_{AB}} \cdot P_{AB} \\
&= -\frac{E_{AB} A_{AB}}{l_{AB}^3} x_{AB}(x_{AB} u_A + y_{AB} v_A + z_{AB} w_A) \\
Y_{A(AB)} &= \frac{y_{AB}}{l_{AB}} \cdot P_{AB} \\
&= -\frac{E_{AB} A_{AB}}{l_{AB}^3} y_{AB}(x_{AB} u_A + y_{AB} v_A + z_{AB} w_A) \\
Z_{A(AB)} &= \frac{z_{AB}}{l_{AB}} \cdot P_{AB} \\
&= -\frac{E_{AB} A_{AB}}{l_{AB}^3} z_{AB}(x_{AB} u_A + y_{AB} v_A + z_{AB} w_A)
\end{aligned}\right\} \quad (69)$$

The components of load applied to the jacks at joint B may be obtained from equations (69) by reversing the signs.

Equations (69) give the solution to the unit problem. If bars AB, AC,..., AR,..., AN are connected to joint A, then by

REDUNDANT TRUSSES

summation we find that the components of load applied to the jacks at joint A due to a displacement u_A are

$$X_A = -u_A \sum_B^N \frac{E_{AR}A_{AR}}{l_{AR}^3} x_{AR}^2$$

$$Y_A = -u_A \sum_B^N \frac{E_{AR}A_{AR}}{l_{AR}^3} x_{AR} y_{AR} \qquad (70)$$

$$Z_A = -u_A \sum_B^N \frac{E_{AR}A_{AR}}{l_{AR}^3} x_{AR} z_{AR}$$

The components of load applied to the jacks at joint R are

$$X_R = u_A \frac{E_{AR}A_{AR}}{l_{AR}^3} x_{AR}^2$$

$$Y_R = u_A \frac{E_{AR}A_{AR}}{l_{AR}^3} x_{AR} y_{AR} \qquad (71)$$

$$Z_R = u_A \frac{E_{AR}A_{AR}}{l_{AR}^3} x_{AR} z_{AR}$$

Similar equations to (70) and (71) can be written down for displacements v_A and w_A.

We illustrate the relaxation method described above by considering again the framework shown in Fig. 70 (p. 98). Taking co-ordinate axes as shown in the diagram, we use equations (70) and (71) to obtain the following "table of operations" (units cm and kN):

Table of Operations for Unit Displacements

	$X_B =$	$Y_B =$	$X_C =$	$Y_C =$	$X_F =$	$Y_F =$	$X_G =$
$u_B = 1$	−252·0	0	83·3	0	0	0	42·7
$v_B = 1$	0	−159·1	0	0	0	111·1	32·0
$u_C = 1$	83·3	0	−126·0	32·0	42·7	−32·0	0
$v_C = 1$	0	0	32·0	−135·1	−32·0	24·0	0
$u_F = 1$	0	0	42·7	−32·0	−252·0	0	83·3
$v_F = 1$	0	111·1	−32·0	24·0	0	−159·1	0
$u_G = 1$	42·7	32·0	0	0	83·3	0	−126·0

BRACED FRAMEWORKS

The symmetry of this table about the diagonal is an example of Maxwell's reciprocal theorem (§2.11). It will be noted that u_A, v_A, u_D, v_D and v_G are not given in the table of operations, since they are necessarily zero. Similarly, we do not consider X_A, Y_A, X_D, Y_D and Y_G, since these forces are not applied to our imaginary screw-jacks, but to real supports capable of sustaining them.

For each of the jack loads which may occur, one particular displacement is most efficacious in relieving it. For example, X_B is most readily diminished by displacement u_B, and the appropriate displacement for relieving any jack load can be found by choosing that displacement which corresponds to the largest coefficient in the column of the table of operations above. If the relaxation solution is to converge fairly rapidly, it is important that the displacement which relieves a given jack load shall not simultaneously introduce a new jack load elsewhere of nearly equal magnitude. u_B as a means of relieving X_B is satisfactory in this respect, since the jack loads set up elsewhere (X_C and X_G) are less than one-third of the magnitude of the change in X_B. v_B as a means of relieving Y_B is not satisfactory, since the new jack load Y_F is $111 \cdot 1/159 \cdot 1 = 70$ per cent of the change in Y_B. The difficulty can be overcome by combining the displacements in suitable proportions. Relaxations which involve combinations of joint displacements are known as "group relaxations", and the combination of displacements is known as a "group displacement". The speed with which the relaxation solution of a framework can be obtained will often depend upon skilful choice of group displacements to hasten the convergence. Suitable group displacements for the present problem are shown in the table below (units cm and kN):

Table of Operations for Group Displacements

	$X_B =$	$Y_B =$	$X_C =$	$Y_C =$	$X_F =$	$Y_F =$	$X_G =$
$v_B = 1$, $v_F = 1\cdot5$	0	7·5	−48·0	36·0	0	−127·5	32·0
$v_B = 1\cdot5$, $v_F = 1$	0	−127·5	−32·0	24·0	0	7·5	48·0
$u_B = 0\cdot2$, $u_C = 1$	32·9	0	−109·3	32·0	42·7	−32·0	85·0
$u_F = 0\cdot2$, $u_G = 1$	42·7	32·0	8·5	−6·4	32·9	0	−109·3

The operations which we shall use will be the displacements u_B, v_C and u_F together with the group displacements. For none of these operations does the relief of a given jack load introduce a jack load elsewhere greater than 40 per cent of that relieved.

The initial loads on the jacks are $Y_B = 30.0$kN due to the external load, and $Y_C = -(0.3/180) \times 20,000 = -33.3$kN due to the error in length of CG. The relaxation procedure is shown in the table below. At each stage the largest jack load is eliminated by a suitable displacement or group displacement, the additional jack loads caused by the displacement are tabulated, and the new jack loads found by adding the additional jack loads to the previous values. The procedure is repeated until the residual jack loads are negligible (1 per cent of the initial values).

The total displacements of the joints can now be obtained by summing the changes given in the first column of the relaxation table. We find $u_B = -0.021$, $v_B = 0.397$, $u_C = -0.127$, $v_C = -0.243$, $u_F = 0.048$, $v_F = 0.265$, $u_G = 0.124$cm. A useful check on arithmetic is obtained by applying these values to the table of operations and making sure that the jack loads are eliminated. If small errors have occurred, so that some jack loads remain, these displacements can be used as the starting-point of a new solution. It is one of the virtues of the relaxation method that arithmetic errors are not of great importance, since any rough solution of the problem can be used to begin a new relaxation.

Once the joint displacements are known, the elongations of the members can be found from equation (23), and the forces in them can be determined. In the present problem, for example, $\delta l_{BF} = v_F - v_B = 0.132$cm and $\delta l_{CG} = v_G - v_C = 0.243$cm. The force in bar BF, $P_{BF} = -0.132/180 \times 20,000 = -14.7$kN. Bar CG was originally 0.3cm too long, so that

$$P_{CG} = \frac{(0.243 - 0.300)}{180} \times 20,000 = -6.3 \text{kN}$$

These values may be compared with the "exact" solution $P_{BF} = -14.8$kN, $P_{CG} = -6.6$kN found in §3.3. The discrepancies are of the same order as the residual errors in the jack loads.

The relaxation method described above is essentially similar to that of Southwell (1940). In an alternative method due to Baker and Ockleston (1935) the forces set up in the adjoining members due to the release of a given jack load constitute the operations table, and

BRACED FRAMEWORKS

Relaxation Solution for Framework of Fig. 70

Displacement	$X_B =$	$Y_B =$	$X_C =$	$Y_C =$	$X_F =$	$Y_F =$	$X_G =$
All zero		30·0		−33·3			
$v_C = -0.247$	0	0	−7·9	33·3	7·9	−5·9	0
	0	30·0	−7·9	0	7·9	−5·9	0
$v_B = 0.353$							
$v_F = 0.235$	0	−30·0	−7·5	5·6	0	1·8	11·3
	0	0	−15·4	5·6	7·9	−4·1	11·3
$u_B = -0.028$							
$u_C = -0.141$	−4·6	0	15·4	−4·5	−6·0	4·5	−1·2
	−4·6	0	0	1·1	1·9	0·4	10·1
$u_F = 0.018$							
$u_G = 0.092$	3·9	3·0	0·8	−0·6	3·0	0	−10·1
	−0·7	3·0	0·8	0·5	4·9	0·4	0
$u_F = 0.019$	0	0	0·8	−0·6	−4·9	0	1·6
	−0·7	3·0	1·6	−0·1	0	0·4	1·6
$v_B = 0.035$							
$v_F = 0.024$	0	−3·0	−0·8	0·6	0	0·2	1·1
	−0·7	0	0·8	0·5	0	0·6	2·7
$u_F = 0.005$							
$u_G = 0.025$	1·1	0·8	0·2	−0·2	0·8	0	−2·7
	0·4	0·8	1·0	0·3	0·8	0·6	0
$u_B = 0.002$							
$u_C = 0.009$	0·3	0	−1·0	0·3	0·4	−0·3	0·1
	0·7	0·8	0	0·6	1·2	0·3	0·1
$u_F = 0.005$	0	0	0·2	−0·2	−1·2	0	0·4
	0·7	0·8	0·2	0·4	0	0·3	0·5
$v_B = 0.009$							
$v_F = 0.006$	0	−0·8	−0·2	0·2	0	0	0·3
	0·7	0	0	0·6	0	0·3	0·8
$u_F = 0.001$							
$u_G = 0.007$	0·3	0·2	0·1	0	0·2	0	−0·8
	1·0	0·2	0·1	0·6	0·2	0·3	0
$u_B = 0.004$	−1·0	0	0·3	0	0	0	0·1
	0	0·2	0·4	0·6	0·2	0·3	0·1
$v_C = 0.004$	0	0	0·1	−0·6	−0·1	0·1	0
	0	0·2	0·5	0	0·1	0·4	0·1
$u_B = 0.001$							
$u_C = 0.005$	0·2	0	−0·5	0·1	0·2	−0·1	0
Jack loads negligible	0·2	0·2	0	0·1	0·3	0·3	0·1

the relaxation solution proceeds entirely in terms of forces, without direct consideration of displacements. Baker and Ockleston term their analysis a "distribution" method (by analogy with the "moment distribution" method of Hardy Cross (1930)): it is described briefly in Pippard and Baker (1957).

EXAMPLES

3(a). Determine the force in member BC of the truss shown in Fig. 67(a) (p. 90), assuming that the material of construction has a linear stress–strain relationship.

3(b). The truss shown in Fig. 69(a) (p. 94) is subjected to vertical loads of 25kN at D and at D'. Determine the reactions at F and at F' due to these loads.

3(c). Two inextensible bars, each of length l, are pin-connected to each other at A and to abutments as shown in Fig. 80. A is initially a small distance r above the abutments. One abutment is rigid. The other resists horizontal movement through a spring of stiffness k. A vertical load P is applied to the joint A.

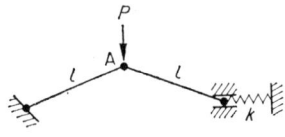

FIG. 80

Determine a relationship between the load P and the vertical deflection v of A. Draw a graph of Pl^2/kr^3 as a function of v/r and show that for positive values of P the system is unstable for values of v/r between 0·422 and 2·000.

3(d). Find the forces in bars BF and CG of the truss shown in Fig. 70(a) (p. 98) by the method of complementary energy.

3(e). Determine the forces in bars FD and DC of the truss shown in Fig. 78(a) (p. 118), when it is subjected to the given load and temperature rises, by the relaxation method.

3(f). The truss shown in Fig. 75(a) (p. 111) is reconstructed with an additional member JF. All bars are free from stress when the truss is unloaded. Find the force in JF due to the loads shown.

3(g). The truss shown in Fig. 78(a) (p. 118) has the support at C replaced by one which is still free to roll horizontally, but which has a vertical deflection given by the equation

$$\delta_v = 10^{-1} V_C + 10^{-2} V_C^2$$

where δ_v is the deflection in cm and V_C is the reaction in kN. Determine the forces in members FD and DC of the truss due to the applied load and temperature rises.

3(h). $4n$ identical elastic bars are all connected to a common joint J. Their further ends are connected to points $A_1 \ldots A_{4n}$ equally spaced around a rigid base

BRACED FRAMEWORKS

circle so that the bars are generators of a right circular cone. By tightening a turnbuckle in bar JA_1, a force P is introduced in that bar. Find the force in bar JA_r.

Show that your answer is independent of the angle of the cone. Show further that the answer is incorrect when the cone degenerates into a planar figure (the bars thus becoming spokes of a wheel) and explain why this should be so.

3(i). Find the vertical displacement of joint D of the truss shown in Fig. 78(a) due to the given load and temperature rises.

3(j). An elastic framework is made in the form of a long ladder. The rungs, which are each of length l, are $0·75l$ apart. Each bay is braced by two diagonal members. The cross-sectional area of a rung is A and of any other member of the framework is $4A$. By means of a turnbuckle, a force P is introduced in an end rung (rung 1). Find the force in rung r.

3(k). The elastic framework shown in Fig. 81 is assembled without pre-stress. All the members have the same cross-sectional area. Find the force in member FG due to the given load.

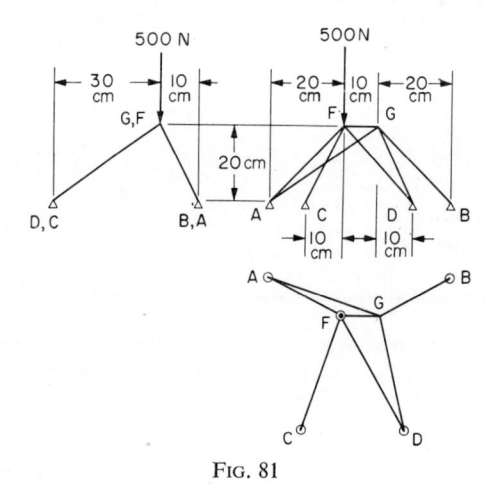

Fig. 81

4. secondary stresses and the failure of braced frameworks

4.1. Analytical Assumptions

In the preceding three chapters, we have assumed in the analyses that our frameworks could be treated as though they were pin-jointed, whether in fact the joints were pinned or not. The reasons for this assumption were discussed in §1.8. As far as the determinations of axial forces in the bars and of deflections of the joints are concerned, the assumption is a good one, as will be demonstrated in §4.3. Real trusses, however, are commonly made with welded or riveted joints which offer a high degree of resistance to relative rotation of the members. Such joints are known as rigid joints, or (if they do not entirely inhibit relative rotation) as semi-rigid joints.

Due to the rigidity of the joints, the members of a braced framework subjected to a change of environment will not only change in length, they will also bend. As we have demonstrated by physical reasoning in §1.8, and as will be shown numerically in §4.3, this bending of the members does not significantly affect the values of the axial forces.† Since the axial forces determine the elongations

---

† The first major investigation of the effects of rigid joints was made by Manderla (1880), but because he did not make use of this present assumption, his equations were too complicated for practical application.

of the bars and these, in turn, determine the displacements of the joints, it follows that these displacements will have almost the same values, whether the joints be pinned or rigid.

We make use of this result, due to Mohr (1892), in determining the amount of bending in the bars of a rigidly jointed framework. The displacements of the joints are first calculated assuming that the members are pinned together, and then these values of the displacements are used to find the bending stresses in the bars of the rigidly jointed frame. The stresses due to bending in a braced framework are known as "secondary stresses". They are usually fairly small compared with the primary stresses due to the axial forces in the members. Once the secondary stresses have been found, it is possible to calculate the error in the primary stresses due to the assumption of pin-joints. This error will usually be found to be so small that no further correction is necessary (see §4.3).

It is possible for there to be bending of the members of a pin-jointed framework, if the centre lines of the members at a joint do not pass through one point. Frameworks of this kind present no particular analytical difficulty. The load–extension characteristics of the eccentrically loaded members are different from those which would obtain if the members were loaded truly axially, but the framework may otherwise be treated as pin-jointed.

4.2. Members Subjected to End Load and Bending

Consider a structural member AB, of length l as shown in Fig. 82. Let the member be subjected to an axial force P and transverse forces F (by resolving laterally the transverse forces may be seen to be equal). Let a clockwise moment M_A be applied to the member at A and a clockwise moment M_B at B. Let the member suffer a lateral deflection at A of y_A and at B of y_B. Then by taking moments about A it will be found that for equilibrium

$$F = -\frac{M_A+M_B}{l}+\frac{P(y_B-y_A)}{l} \qquad (72)$$

Let a coordinate x be measured along AB from A, and let y denote the deflection of the member from its original position. Then

SECONDARY STRESSES AND FAILURE

at section x of the member there is applied an axial force P, a transverse force F and a moment M given by

$$M = M_A + Fx - P(y - y_A) \qquad (73)$$

F is known as the "shear force" at x and M as the "bending moment" at x. Positive bending moment is associated with concave upwards curvature, so that the bending moment at A is M_A, and at

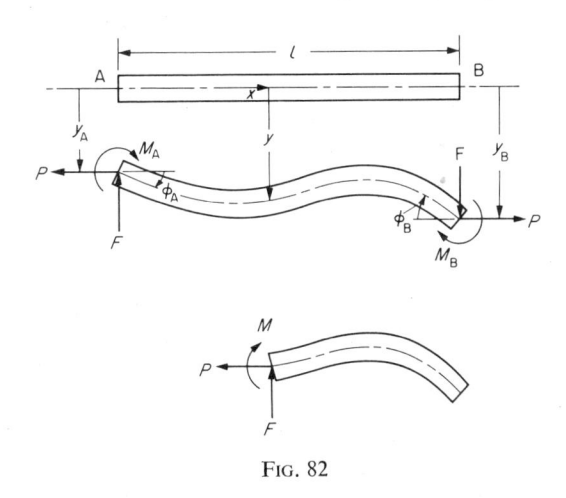

Fig. 82

B is $-M_B$. (It should be particularly noted that the sign convention for the internal bending moment M in a member and the sign convention for the externally applied end moments M_A and M_B give similar signs at the left-hand end of the member but opposite signs at the right-hand end—see also Fig. 87.)

It is found experimentally that the lateral deflections y of the member are dependent almost entirely on the distribution of the bending moment M and that the lateral deflections due to the shear force F and the axial force P are either zero or negligible. It is further found that at any given section x of the member there is a relationship between the bending moment M and the curvature (reciprocal of radius of curvature). For a uniform member this relationship will be the same for all values of x.

The bending moment–curvature relation is not, in general, linear. Fig. 83 shows an example for a 1cm-square bar of mild steel. For a

BRACED FRAMEWORKS

member made from a linear elastic material, the curvature is proportional to the bending moment. We have

$$\text{curvature} = \frac{1}{EI} \times \text{bending moment} \tag{74}$$

where EI is a property of the section known as the "flexural rigidity" or "bending stiffness".

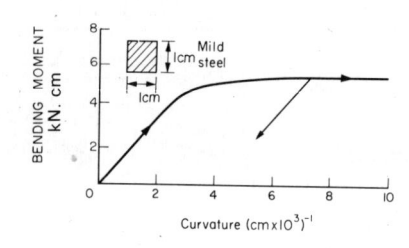

FIG. 83. Bending moment–curvature relationship for mild steel bar

It is shown in textbooks on coordinate geometry that the curvature of any curve defined by Cartesian coordinates x, y is given by

$$\text{curvature} = -\frac{d^2y/dx^2}{\{1+(dy/dx)^2\}^{3/2}} \tag{75}$$

(the minus sign occurs because in equation (74) we have associated positive curvature with positive bending moment—i.e. bending concave upwards, whereas positive d^2y/dx^2 corresponds to bending concave downwards).

Combining equations (72), (73), (74) and (75) we find that

$$M_A - (M_A + M_B)\frac{x}{l} - P\left\{y - y_A + (y_A - y_B)\frac{x}{l}\right\}$$
$$= \text{bending moment}$$
$$= -\frac{EI(d^2y/dx^2)}{\{1+(dy/dx)^2\}^{3/2}} \tag{76}$$

Now in practical braced frameworks the lateral deflections y of the members will be small. It follows that the rotations dy/dx will also

SECONDARY STRESSES AND FAILURE

be small, and $(dy/dx)^2$ will be negligible compared with unity. The denominator of the right-hand side of equation (76) may therefore be replaced by unity. Further, in most members, the term $P\{y-y_A+(y_A-y_B)(x/l)\}$ in the expression for the bending moment will be negligible compared with M_A or M_B (provided that if P is negative it is small compared with the critical value for instability—see §4.5.3). Equation (76) may then be simplified to

$$M_A - (M_A + M_B)\frac{x}{l} = -EI\frac{d^2y}{dx^2} \qquad (77)$$

This equation may be integrated twice to give

$$y = \frac{-M_A x^2}{2EI} + \frac{(M_A + M_B)x^3}{6EIl} + Cx + D$$

where C and D are arbitrary constants of integration. C and D may be found from the conditions that $y = y_A$ and y_B at $x = 0$ and l. We then obtain

$$y = \frac{-M_A x^2}{2EI} + \frac{(M_A + M_B)x^3}{6EIl} + \frac{(2M_A - M_B)lx}{6EI} + y_A + (y_B - y_A)\frac{x}{l}$$

By differentiating this equation we obtain the slopes $\phi\ (=dy/dx)$ of the member at A and B as

$$\left.\begin{aligned}\phi_A &= \frac{l}{6EI}(2M_A - M_B) + \frac{1}{l}(y_B - y_A) \\ \phi_B &= \frac{l}{6EI}(2M_B - M_A) + \frac{1}{l}(y_B - y_A)\end{aligned}\right\} \qquad (78)$$

These equations can be solved to give

$$\left.\begin{aligned}M_A &= \frac{2EI}{l}(2\phi_A + \phi_B) + \frac{6EI}{l^2}(y_A - y_B) \\ M_B &= \frac{2EI}{l}(2\phi_B + \phi_A) + \frac{6EI}{l^2}(y_A - y_B)\end{aligned}\right\} \qquad (79)$$

We shall show in the next section how equations (78) or (79) can be used to determine the secondary stresses in rigidly jointed plane

BRACED FRAMEWORKS

trusses. The first systematic use of these equations was made by Bendixen (1914).

4.3. Solution by Slope-deflection Equations

The determination of the bending stresses in rigidly jointed plane trusses by direct use of the slope-deflection equations (78) or (79) can best be demonstrated by an example.

Consider the truss shown in Fig. 84. The members are all made from the same uniform section (207mm × 134mm × 30kg Universal

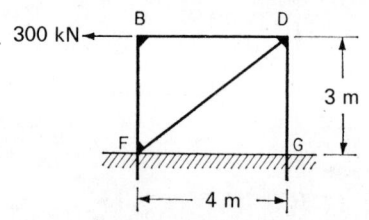

$A = 38$ cm^2
$E = 210$ GN/m^2
$EI = 60.5$ GN.cm^2
$Z = 279$ cm^3

Fig. 84

Beam). If we regard the truss for the present as pin-jointed, the forces in the members can readily be obtained as $P_{BD} = 300$, $P_{BF} = 0$, $P_{DF} = -375$ and $P_{DG} = 225$kN. By calculating the elongations of the members and drawing a displacement diagram, the displacements of the joints will be found to be as follows:

B, to the left   0·508cm
up   0
D, to the left   0·357cm
up   0·085cm
perpendicular to DF   0·282cm

Now consider the rigidly jointed frame. Let the clockwise moment applied to member BD at B be M_{BD} and at D be M_{DB}, and so on. Let the clockwise rotations of joints B and D be ϕ_B and ϕ_D (ϕ_F and ϕ_G are zero). Then if we assume that the displacements of the joints

SECONDARY STRESSES AND FAILURE

are given correctly by the analysis of the pin-jointed frame, these displacements may be substituted in equations (79) to give relations between the moments and rotations as follows:

$$\left.\begin{aligned}
M_{BD} &= \frac{2EI}{400}(2\phi_B + \phi_D) + \frac{6EI}{400^2}(0 + 0.085) \\
M_{DB} &= \frac{2EI}{400}(2\phi_D + \phi_B) + \frac{6EI}{400^2}(0 + 0.085) \\
M_{BF} &= \frac{2EI}{300}(2\phi_B + 0) + \frac{6EI}{300^2}(0.508 - 0) \\
M_{FB} &= \frac{2EI}{300}(0 + \phi_B) + \frac{6EI}{300^2}(0.508 - 0) \\
M_{DG} &= \frac{2EI}{300}(2\phi_D + 0) + \frac{6EI}{300^2}(0.357 - 0) \\
M_{GD} &= \frac{2EI}{300}(0 + \phi_D) + \frac{6EI}{300^2}(0.357 - 0) \\
M_{FD} &= \frac{2EI}{500}(0 + \phi_D) + \frac{6EI}{500^2}(0 + 0.282) \\
M_{DF} &= \frac{2EI}{500}(2\phi_D + 0) + \frac{6EI}{500^2}(0 + 0.282)
\end{aligned}\right\} \quad (80)$$

At joints B and D, the sum of the moments applied to the members must be zero, since no external moment is applied to the joint. We therefore have as equations of equilibrium,

$$\left.\begin{aligned} M_{BD} + M_{BF} &= 0 \\ M_{DB} + M_{DF} + M_{DG} &= 0 \end{aligned}\right\} \quad (81)$$

Substituting equations (80) in equations (81) we obtain two simultaneous equations for ϕ_B and ϕ_D, from which

$$\phi_B = -0.00140, \quad \phi_D = -0.00085$$

These values may now be substituted back in equations (80) in order

BRACED FRAMEWORKS

to determine the bending moments. We find, on putting EI equal to 60.5 GN.cm^2,

$$M_{BD} = -9.1 \text{kN.m} \qquad M_{DG} = 7.5 \text{kN.m}$$
$$M_{DB} = -7.5 \qquad\qquad M_{GD} = 10.9$$
$$M_{BF} = 9.1 \qquad\qquad M_{FD} = 2.0$$
$$M_{FB} = 14.8 \qquad\qquad M_{DF} = -0.0$$

In a linear elastic member, the maximum bending stress can be obtained by dividing the bending moment by a property of the section known as the "section modulus", Z. If the section is symmetrical, the maximum bending stress has equal tension and compression values on opposite sides of the member. For the present section, Z is equal to 279cm^3. The member carrying the largest axial (primary) stress is DF, with a stress of $-375/38 = -9.9$kN/cm^2. The largest bending moment in DF occurs at F and is equal to 2.0kN.m. The bending stress (secondary stress) at F is thus $200/279 = \pm 0.7$kN/cm^2. It will be seen that the secondary stress due to bending is about 7 per cent of the primary stress due to axial force.

It is of interest to consider the values of the shear forces in the members, since these represent additional forces at the joints which have been ignored in the primary stress analysis. From equations (72), ignoring the small term $P(y_B - y_A)/l$, we have

$$F_{BD} = -\frac{(-9.1-7.5)}{4} = 4.2 \text{kN}$$

$$F_{BF} = -\frac{(9.1+14.8)}{3} = 8.0 \text{kN}$$

$$F_{DG} = -\frac{(7.5+10.9))}{3} = -6.1 \text{kN}$$

$$F_{FD} = -\frac{(2.0-0.0)}{5} = -0.4 \text{kN}$$

It will be seen that the largest force (F_{BF}) is less than 3 per cent of the value of the external load. The errors in the axial forces due to ignoring the shear forces (i.e. ignoring the rigidity of the joints) will therefore be small, and our assumption that the axial forces and the displacements of the joints can be obtained with sufficient accuracy by treating the frame as though it were pin-jointed is justified.

The method of solution described above, in which simultaneous equations for the joint rotations ϕ are set up, will usually be found convenient provided the number of joints is small, although the rotations can be eliminated by using equations (78) instead of (79) and then solving directly for the bending moments.

Semi-rigid joints present no particular analytical difficulty, but we shall not consider them here. Their behaviour is described by Pippard and Baker (1957).

Inelastic members can also be treated quite simply. Since the bending stresses are small compared with the axial stresses, the flexural rigidity must be calculated using a modulus equal to the slope of the stress–strain curve of the material at the appropriate value of the primary (axial) stress.

Solution by direct use of the slope-deflection equations is satisfactory provided the number of joints is small. For larger frames, the number of simultaneous equations to be solved becomes prohibitive and as in the case of the determination of axial forces (§3.6) we turn to relaxation methods.

4.4. Solution by Moment Distribution

The determination of the bending moments in a rigidly jointed frame by moment distribution (Hardy Cross, 1930) is precisely analogous to the determination of the axial forces by Southwell's successive relaxation of constraints (§3.6). The method was first applied to the determination of secondary stress by Thompson and Cutler (1932). The "unit problem" in the present case consists of determining the moments set up in the members connected to it when one joint of the frame is subjected to a moment M. This joint is supposed free to rotate, but cannot be displaced. All other joints of the frame are rigidly held. Let the joint be A (Fig. 85) and let it be connected by uniform elastic members AB, AC,..., AN to joints B, C,..., N. Suppose that the rotation of joint A is ϕ_A. Then from equations (79), noting that all displacements are zero and all

BRACED FRAMEWORKS

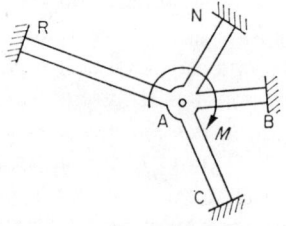

FIG. 85

rotations except ϕ_A are zero, we have for member AR (using the notation of the previous section),

$$M_{AR} = \frac{4E_{AR} I_{AR}}{l_{AR}} \phi_A \\ M_{RA} = \frac{2E_{AR} I_{AR}}{l_{AR}} \phi_A = \tfrac{1}{2} M_{AR}$$ (82)

Summing all the moments applied to the ends A of the members, for equilibrium,

$$M = \sum_B^N M_{AR} = \sum_B^N \frac{4E_{AR} I_{AR}}{l_{AR}} \phi_A$$

whence
$$\phi_A = \frac{M}{\sum_B^N \frac{4E_{AR} I_{AR}}{l_{AR}}}$$ (83)

Substituting equation (83) in the first of equations (82),

$$M_{AR} = \frac{(E_{AR} I_{AR})/l_{AR}}{\sum_B^N (E_{AR} I_{AR})/l_{AR}} \cdot M$$ (84)

If we apply a moment M to joint A it will be distributed among the members according to equation (84), and there will be applied at the far end of each member a moment equal to one-half of that at end A, in accordance with the second of equations (82). The factor preceding M in equation (84) is known as the "moment distribution factor" for member AR at joint A.

We have now solved the unit problem. We must next consider how this solution is to be used. Let us suppose that we have a rigidly jointed frame subjected to given changes of environment. Let the

SECONDARY STRESSES AND FAILURE

joints move to those positions determined by the primary stress analysis, but let rotation of the joints be completely prevented. Then there will be a moment applied at each end of every member. These moments are known as the "fixed-end moments", i.e. the moments set up in the members due to the fact that their ends cannot rotate. They can be determined from equations (79), putting $\phi_A = \phi_B = 0$ and using the values of y_A and y_B from the primary stress analysis. The sum of the fixed-end moments at any given joint will not, in general, be zero, so that we have to imagine some form of external jack providing the "residual" moment. These residual moments can be calculated for each joint of the frame.

The method of solution by moment distribution consists of relaxing the jack carrying the largest residual moment and distributing this moment among the members at that joint in accordance with equation (84). It will be noted that moments appear at the far end of each member equal to one-half of those at the joint concerned (equation (82)). The factor one-half is known as the "carry over factor". After the first relaxation has been completed, the residual moments can be re-totalled and the jack which now carries the largest residual moment can be relaxed. The process is repeated until the residual moments are negligible. Since the carry over factor is one-half, the process is fairly rapidly convergent.

We shall illustrate the method by considering again the truss shown in Fig. 84. The distribution factors at joint B are

$$\frac{EI/3}{EI/3+EI/4} : \frac{EI/4}{EI/3+EI/4} = 0.572 : 0.428$$

and at joint D are

$$\frac{EI/4}{EI/4+EI/5+EI/3} : \frac{EI/5}{EI/4+EI/5+EI/3} : \frac{EI/3}{EI/4+EI/5+EI/3} = 0.319 : 0.255 : 0.426$$

We are not interested in the distribution factors at F and G, since these joints cannot rotate. The displacements of the joints are given in §4.3. We use these in equations (79) to obtain the fixed-end moments as $M_{BF} = M_{FB} = 20.5$, $M_{BD} = M_{DB} = 1.9$, $M_{FD} = M_{DF} = 4.1$, $M_{DG} = M_{GD} = 14.4$ kN.m.

BRACED FRAMEWORKS

The moment distribution process is shown in Fig. 86, where the fixed-end moments at each joint are enclosed by broken lines. The joint which initially has the largest residual moment is B. This residual moment is equal to $20.5+1.9 = 22.4$kN.m. To relax the jack at B we must apply a moment of -22.4kN.m to the joint. This is divided between members BF and BD in the ratio of

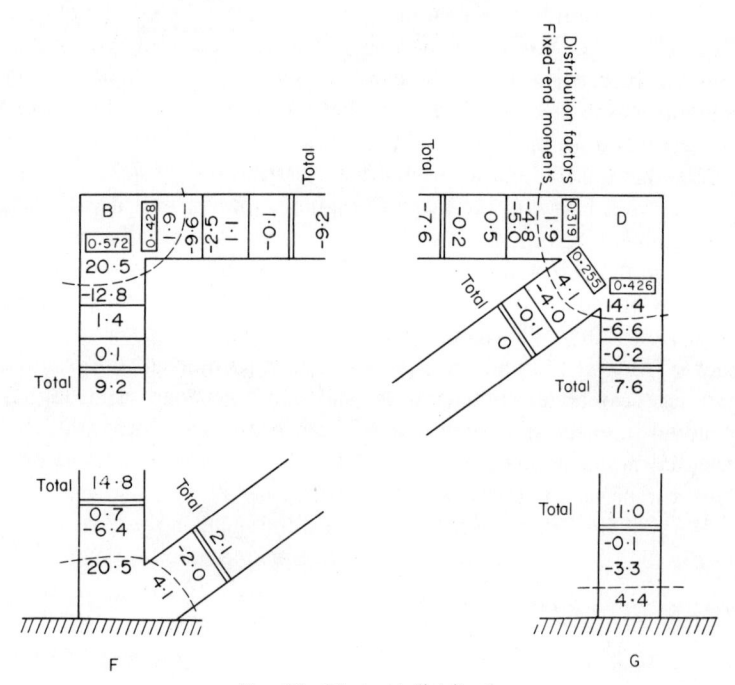

FIG. 86. Moment distribution

the distribution factors: $0.572 \times -22.4 = -12.8$ to member BF, and $0.428 \times -22.4 = -9.6$ to BD. When these values have been inserted, they are underlined to show that at this stage the joint is balanced. Since the "carry over factor" is one-half, a moment of $\frac{1}{2} \times -12.8 = -6.4$kN.m must be applied to BF at F, and a moment of $\frac{1}{2} \times -9.6 = -4.8$kN.m to BD at D.

We now consider joint D. The residual moment is $1.9+4.1+14.4-4.8 = 15.6$kN.m. To relax the jack at B, moments of 0.319, 0.255 and 0.426×-15.6 must be applied to the members. These may be evaluated as $-5 \cdot 0$, $-4 \cdot 0$ and $-6 \cdot 6$kN.m applied to DB, DF and DG. These moments are underlined to show that the joint is balanced. Carry over moments of one-half of these values occur at B, F and G. Joint B is thus unbalanced and must be relaxed again. The process is repeated until the residual moments are negligible (0·1kN.m or less).

The final moments at the ends of each member can now be obtained by summation. The fact that the moments at each joint should total zero provides some check on the arithmetic. The moments shown in Fig. 86 may be compared with those obtained in §4.3 by using the slope-deflection equations: it will be seen that the differences are negligible.

The moment distribution method can be used for members having a pin-joint at one end provided $\tfrac{3}{4}EI/l$ is used instead of EI/l in equation (84) and provided the carry over factor is taken as zero. The reader may like to show this for himself by using the first of equations (78). The method can also be used for semi-rigid joints and for non-prismatic members, when the carry over factor will not generally be one-half. Inelastic members can be treated in a similar manner to that used for the slope-deflection equations.

4.5. The Failure of Braced Frameworks

4.5.1. *Types of failure.* It was stated in §1.1 that failure of a structure occurs when the deformations exceed certain prescribed limits. Failure may be due to a single set of environmental changes, or it may occur after a given set of environmental changes has been repeated a number of times. Failures due to a single set of environmental changes include those in which the load was too large, the temperature too high or the time too long (creep). Failures due to a number of cycles of environmental changes include fatigue, fretting and incremental collapse.

Failure may be primarily due to the weakness of a given member or of a given joint, or it may be associated with the behaviour of the whole framework. Examples of the former type include tensile fracture of bars and shearing of rivets: the latter include such phenomena as the lateral instability of bridge trusses.

We cannot hope, in a book of the present length, to explore all of these types of failure in detail. All that we can do is to display some of the commoner and simpler forms. Before doing so, we must consider some widely used criteria of structural safety.

4.5.2. *Criteria of safety*. In many structural systems it is the intention of the designer that all parts shall behave elastically at all times. The deformations associated with the elastic stresses are regarded as of acceptable magnitude and no permanent deformations of the system can occur. The criterion of safety (we use the word "safety" here to mean "satisfactory function") is that the stress shall everywhere be below the elastic limit of the material: the elastic limit is defined as that stress at which the stress-strain curve for the material ceases to be linear. In practice the elastic limit is difficult to determine and we use instead the yield stress, for a material exhibiting a well-defined yield, or the proof stress, for a material having a smooth stress-strain curve. The proof stress is defined as that stress which will produce a small permanent strain (usually taken as 0·1 per cent or 0·2 per cent) on unloading. A factor of safety n is incorporated in the criterion, so that we have finally

$$\text{stress} < \frac{1}{n} \times \text{yield stress} \tag{85}$$

(n is commonly about 1·7).

This way of defining satisfactory function is known as the "stress factor" method. The method is very widely used, and structures which obey this criterion are usually safe, but the method is open to a number of objections. The precise elastic analysis of every part of a structure is often not possible, either due to its complexity or to uncertainty about such matters as differential settlement of foundations or the fit of joints. It follows that some parts of almost all structures disobey equation (85) and the method is invalid in detail. A system which satisfies equation (85) at all points may still fail unstably due to a small disturbance. A further criticism of the stress factor method is that in redundant systems, elastic design can be very inefficient.

An alternative criterion of safety is the "load factor" method. Here the load which would cause failure of the structure is calculated and the actual load is restricted to some smaller value, so that

$$\text{load} < \frac{1}{n} \times \text{failing load} \tag{86}$$

SECONDARY STRESSES AND FAILURE

where n is the load factor. This method is generally more satisfactory than the stress factor method, but it can lead to incremental types of failure in structures subjected to cycles of loading.

In some structural systems, such as aircraft wings, suspension bridges or control cables, the deflections may need to be limited to values smaller than those given by the strength requirements of equations (85) and (86). It may also be necessary to maintain high stiffness in a structure for reasons of stability. Under these circumstances a stiffness criterion is used.

In structures subjected to cyclic changes of environment, or in which creep is significant, the deformations may become unacceptable after a certain time. Cyclic changes of stress may also produce fatigue failure after a limited period. For systems of this kind a "safe life" criterion is used in which the structure or parts of it must be replaced after a period of service. The "safe life" method may be combined with a "fail safe" philosophy of design, in which the structure is so arranged that failure of any one part does not cause catastrophic failure of the whole system.

In the following sections we shall see how some of these criteria of safety may be applied to complete structural systems or to parts of them.

4.5.3. *Failure of an individual member.* Consider a uniform elastic member, initially straight, and suppose that the axial force in the member and the moments applied at its ends are known. Then from equation (76), regarding $(dy/dx)^2$ as negligible compared with unity, but allowing for the terms in P which we have previously neglected, we have

$$EI\frac{d^2y}{dx^2} - Py = \{M_A + M_B + P(y_A - y_B)\}\frac{x}{l} - M_A - Py_A \quad (87)$$

The solution of this equation may be written as

$$y = C\sinh\sqrt{\frac{P}{EI}}\cdot x + D\cosh\sqrt{\frac{P}{EI}}\cdot x - \left\{\frac{M_A + M_B}{P} + (y_A - y_B)\right\}\frac{x}{l} + \frac{M_A}{P} + y_A$$

where C and D are arbitrary constants of integration. C and D may

BRACED FRAMEWORKS

be found from the conditions $y = y_A$ at $x = 0$ and $y = y_B$ at $x = l$. We then have

$$y = \left(\frac{M_B}{P} + \frac{M_A}{P}\cosh\sqrt{\frac{P}{EI}}\cdot l\right)\frac{\sinh\sqrt{(P/EI)}x}{\sinh\sqrt{(P/EI)}l} +$$
$$+ \frac{M_A}{P}\left(1 - \cosh\sqrt{\frac{P}{EI}}\cdot x\right) - \left\{\frac{M_A + M_B}{P} + (y_A - y_B)\right\}\frac{x}{l} + y_A$$

The bending moment at any section x,

$$-EI\frac{d^2y}{dx^2} = \frac{1}{\sinh\sqrt{(P/EI)}l}\left\{M_A\sinh\sqrt{\frac{P}{EI}}\cdot(l-x) - M_B\sinh\sqrt{\frac{P}{EI}}\cdot x\right\} \quad (88)$$

To find the section where the bending moment is greatest, we must consider a number of separate cases. If P is positive, the maximum bending moment will occur at one end of the member, irrespective

P positive
M_A and M_B of same sign

P positive
M_A and M_B of opposite sign

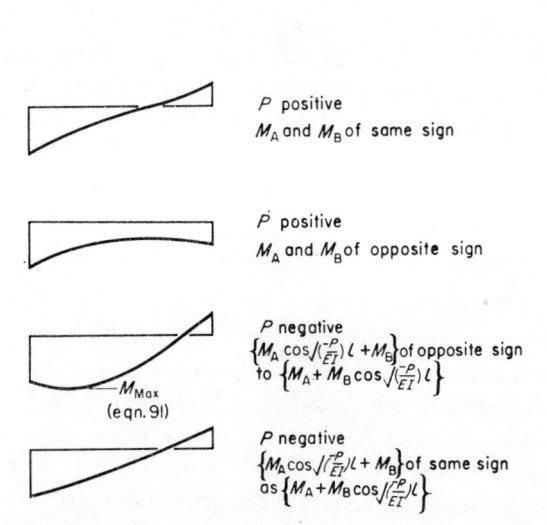

P negative
$\{M_A\cos\sqrt{(\frac{-P}{EI})}l + M_B\}$ of opposite sign
to $\{M_A + M_B\cos\sqrt{(\frac{-P}{EI})}l\}$
M_{Max} (eqn. 91)

P negative
$\{M_A\cos\sqrt{(\frac{-P}{EI})}l + M_B\}$ of same sign
as $\{M_A + M_B\cos\sqrt{(\frac{-P}{EI})}l\}$

FIG. 87. Bending moment diagrams (it should be noted that the bending moment at A is M_A and at B is $-M_B$)

SECONDARY STRESSES AND FAILURE

of the signs of M_A and M_B (Fig. 87). If P is negative, equation (88) can be rewritten as

$$-EI\frac{d^2y}{dx^2} = \frac{1}{\sin\sqrt{(-P/EI)}\,l}\left\{M_A \sin\sqrt{\frac{-P}{EI}}\cdot(l-x) - M_B \sin\sqrt{\frac{-P}{EI}}\cdot x\right\} \quad (89)$$

The bending moment will have its greatest numerical value at a section other than the ends provided the slopes of the bending moment diagram at $x=0$ and $x=l$ are of opposite sign. By differentiating equation (89) we obtain this condition as

$$\left\{M_A \cos\sqrt{\frac{-P}{EI}}\cdot l + M_B\right\} \text{ of opposite sign to}$$

$$\left\{M_A + M_B \cos\sqrt{\frac{-P}{EI}}\cdot l\right\} \quad (90)$$

If the inequality (90) is satisfied, the maximum bending moment can readily be found from equation (89) to be

$$M_{max} = \left\{(M_A^2 + M_B^2)\operatorname{cosec}^2\sqrt{\frac{-P}{EI}}\cdot l + 2M_A M_B \operatorname{cosec}\sqrt{\frac{-P}{EI}}\cdot l \cot\sqrt{\frac{-P}{EI}}\cdot l\right\}^{\frac{1}{2}} \quad (91)$$

If inequality (90) is not satisfied, the maximum bending moment occurs at one end (Fig. 87).

Once the maximum bending moment in the member has been determined, the maximum stress can be obtained as

$$\sigma_{max} = \frac{P}{A} \pm \frac{M_{max}}{Z} \quad (92)$$

If the stress factor criterion of safety is used, σ_{max} must be less than the value given by equation (85).

Provided the axial force and end moments applied to an individual member are known, the analysis given above provides an accurate means of determining the maximum stress. In many practical applications, however, although the axial force in the member may be known (from the primary stress analysis), the values of the end moments are less certain. This uncertainty may be due to errors associated with ignoring the P term in equation (76), or it may be due to lack of information about the behaviour of the joints: in many cases the uncertainty will occur because time does not permit a secondary stress analysis to be carried out. Further, the analysis above has been developed for members which are initially perfectly straight, whereas practical members may have unknown initial curvatures. For these reasons, certain simplifying assumptions are commonly used in considering the functioning of individual members in braced frameworks.

For members in tension, the end moments and any initial curvature are ignored, and the maximum stress is taken as equal to P/A. This assumption is satisfactory for members made from ductile materials, since bending stresses can be relieved by local yielding.

For members in compression (struts) an analysis due to Perry (1886) is used. Consider a pin-ended member ($M_A = M_B = 0$) having an initial deflection when unloaded of

$$y_0 = c \sin \frac{\pi x}{l}$$

Then the bending moment, which is equal to the flexural rigidity multiplied by the *change* of curvature, is given by

$$-EI\left(\frac{d^2 y}{dx^2} - \frac{d^2 y_0}{dx^2}\right) = -Py$$

Substituting for y_0 and noting that $y = 0$ when $x = 0$ and l, this equation can be solved as

$$y = \frac{c \sin(\pi x/l)}{1 + (l^2 P/\pi^2 EI)}$$

The maximum deflection occurs when $x = \tfrac{1}{2}l$ and the maximum bending moment is thus

$$-Py_{max} = -\frac{Pc}{1 + (l^2 P/\pi^2 EI)}$$

The maximum compressive stress (remembering that P is negative) is

$$\sigma_{max} = \frac{P}{A} + \frac{Pc}{Z\{1 + (l^2 P/\pi^2 EI)\}} \tag{93}$$

SECONDARY STRESSES AND FAILURE

If we write σ_m for P/A, the mean stress, and k for $\sqrt{(I/A)}$, the radius of gyration of the cross-section, equation (93) becomes

$$\sigma_{max} = \sigma_m \left\{ 1 + \frac{cA}{Z} \cdot \frac{1}{1 + (\sigma_m/\pi^2 E)(l/k)^2} \right\} \quad (94)$$

Experiments by Robertson (1925) suggested that in practical pin-ended compression members cA/Z, which is a measure of the initial imperfections, might safely be approximated by $0.003l/k$. Once values for σ_{max} and E have been chosen, equation (94) then becomes an equation between the permissible mean stress in the member σ_m

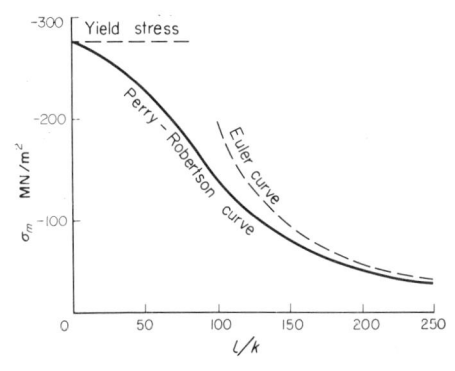

FIG. 88. The Perry–Robertson curve for mild steel

and the "slenderness ratio", l/k. For mild steel, choosing σ_{max} as the yield stress $-280\,\text{MN/m}^2$ and E as $200\,\text{GN/m}^2$, the relationship between σ_m and l/k, which is known as the Perry–Robertson curve, is plotted in Fig. 88.

It will be noted that at low values of the slenderness ratio, the permissible mean stress σ_m is close to the yield stress, as was first pointed out by Lamarle (1846). We should expect this result, since in a short thick member bending effects will be unimportant. At high values of l/k (long slender members) the Perry–Robertson curve is asymptotic to another relationship between σ_m and l/k known as the Euler curve. We can obtain the Euler curve by considering equation (93). As P tends to the value $-\pi^2 EI/l^2$, the maximum compressive stress tends to infinity. It follows that the member will fail at a value of P smaller than this critical value. If we divide by

BRACED FRAMEWORKS

the cross-sectional area A, we obtain the critical mean stress for instability (when the stresses and deflections tend to become infinite) as

$$\sigma_{m.cr} = \frac{-\pi^2 E}{(l/k)^2} \tag{95}$$

$\sigma_{m.cr}$ is termed the Euler critical stress after the Swiss mathematician (Euler, 1744).

It is usual to apply a factor of safety to the value of σ_m obtained from the Perry–Robertson formula (94). If we regard the member as having failed when the yield stress is reached, this factor is essentially a load factor. If, on the other hand, a factor is applied to σ_{max}, the factor is a stress factor.

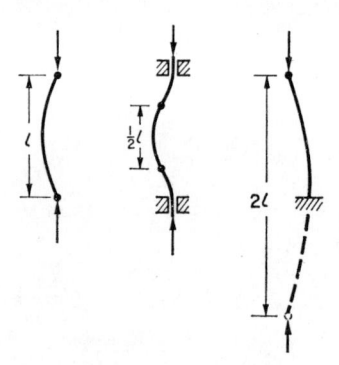

FIG. 89. Effective lengths of compression members (lengths of equivalent pin-ended members)

The Perry–Robertson method described above provides a satisfactory means of assessing the safety of pin-ended compression members. For members having other end conditions, the concept of an "effective length" is used. A member of length l whose ends are completely restrained against rotation has points of contraflexure at sections $\frac{1}{4}l$ and $\frac{3}{4}l$ from one end (Fig. 89). It is therefore regarded as equivalent to a pin-ended member of length $\frac{1}{2}l$, and the Perry–Robertson method can again be used. A member of length l which is fixed at one end and free at the other has an effective length of $2l$ (Fig. 89).

A member of a braced framework which is connected through rigid joints to other members at each end is supposed to have its ends

SECONDARY STRESSES AND FAILURE

partially restrained against rotation. Its effective length thus lies between that for pinned ends (l) and that for ends completely restrained against rotation ($\frac{1}{2}l$): it is usually taken as $0·7l$. This last piece of reasoning is of doubtful validity, since the moments transmitted to the member at its ends (which really determine the points of contraflexure and thus the effective length) depend upon the loading and behaviour of every other member in the structure. Nevertheless, in the absence of more accurate information, an effective length of $0·7l$ is commonly used in the analysis of compression members in rigidly jointed braced frameworks.

So far in this section we have considered the behaviour of elastic members and have supposed failure to occur when the yield stress is reached at one point. If the members are allowed to become plastic, higher loads may be sustained. The behaviour of elasto-plastic members is discussed by Baker, Horne and Heyman (1956). If the material may creep under stress, compression members fail after a limited time: instead of a critical stress we have a critical time. Failure due to creep is discussed by Hoff (1962) and by Odqvist (1962). For members subjected to cyclic forces, fatigue may be important. A general consideration of fatigue is given by Grover, Gordon and Jackson (1956).

4.5.4. *Failure of a joint.* The members of a braced framework are usually connected to one another via a gusset plate (Fig. 17(a)) (p. 28) to which they are either riveted or welded. Failure of the joint may occur due to tearing of the gusset plate or the member in tension or shear, or it may be due to shearing of rivets or welds or elongation of rivet holes (bearing). The analysis of the force carried by a rivet or an element of weld follows one of two methods. The first leads essentially to a stress factor criterion and the second to a load factor criterion.

In the first method, the member and the gusset plate are assumed to be rigid and the force in a rivet (or an element of weld) is supposed to be proportional to the relative movement of the member and gusset plate at the point concerned and to act in the direction of that movement. Thus (Fig. 90(a)) if there are n similar rivets and the member and plate suffer a simple translation, the forces in all rivets will be equal and will act in the same direction. These forces will be statically equivalent to a single force acting through the centre of gravity G of the rivet group, and so a force between member and

BRACED FRAMEWORKS

gusset plate which acts through G will produce equal forces in all rivets.

Now let the member and gusset plate suffer relative rotation about G and let a typical rivet be at (r, θ) from G (Fig. 90(b)). Then the

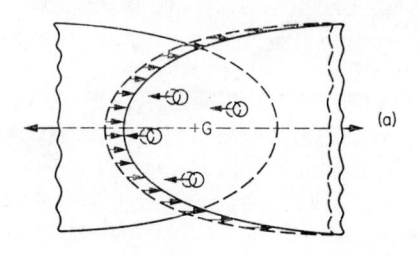

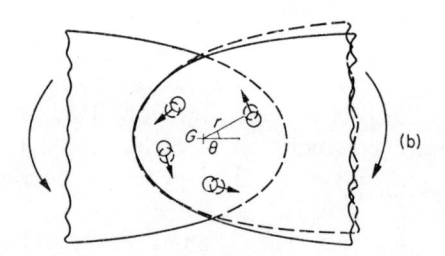

FIG. 90. Forces in rivets or elements of weld

relative displacement and thus the force in the rivet will be proportional to r. Let the force in the rivet be kr: it acts in a direction perpendicular to the radius r. Then the horizontal component of all the rivet forces will be

$$k \sum_1^n r \sin \theta$$

which is equal to zero since G is the centre of gravity of the rivet group. Similarly, the vertical component of the rivet forces,

$$k \sum_1^n r \cos \theta$$

is zero. The rivet forces are thus equivalent to a pure moment whose value we can find by considering moments of the rivet forces about G as

$$M = k \sum_1^n r^2$$

SECONDARY STRESSES AND FAILURE

The force in a typical rivet due to a pure moment M between member and gusset plate is thus

$$F = kr = \frac{Mr}{\sum_{1}^{n} r^2} \qquad (96)$$

The force acts in a direction perpendicular to the line joining the rivet to the centre of gravity of the rivet group.

Any general loading between member and gusset plate can be resolved into a force acting through the centre of gravity of the rivet group or weld, and a moment about that point. The force in a rivet or element of weld consists of two components, one due to the applied force and the other due to the applied moment. These components must be added vectorially to obtain the total force.

As an example we may consider the joint shown in Fig. 91(a). The 10kN applied force between member and gusset plate may be regarded as equivalent to 10kN acting through G together with a moment about G of $10 \times 3.5 = 35$kN.cm. Due to the applied force, each rivet carries a horizontal force of $10/5 = 2$kN. The determination of the rivet forces due to the applied moment is shown in the table of Fig. 91(b). In Fig. 91(c) we show the vectorial addition of the rivet forces due to the applied force and the applied moment to give the total values.

In the second method of determining rivet or weld forces, each rivet or element of weld is assumed to behave plastically. In a braced framework, quite small rotations of the ends of the members are sufficient to relieve the end moments, and it is assumed therefore that the moment acting on the rivet group is removed by plastic yielding of the rivets and that the rivet group is subjected only to a force acting through its centre of gravity. This applied force produces equal forces in all rivets. Unless the centre of gravity of the rivet group coincides with the centre of area of cross-section of the member, this assumption implies that the member will be eccentrically loaded.

Once the force to be carried by a rivet or an element of weld has been determined, this can be compared with an allowable force based upon experiment. The shear force allowed in a rivet is usually obtained from a (fictitious) allowable shear stress multiplied by the area of cross-section: for mild steel rivets the allowable shear stress

BRACED FRAMEWORKS

will be about 9kN/cm². The shear force allowed in an element of weld is taken as the allowable shear stress multiplied by the area of the "throat" (see Fig. 92). The bearing force allowed in the member or gusset plate is determined from a (fictitious) allowable bearing

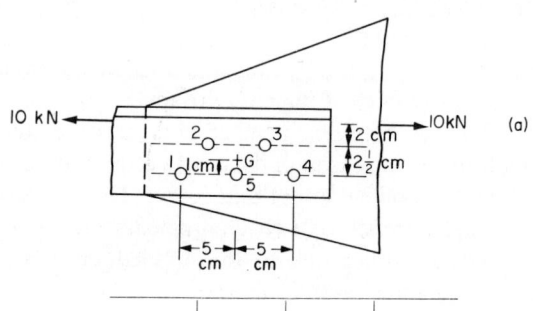

Rivet	r	r^2	$Mr/\Sigma r^2$
1	5·099	26·0	2·550
2	2·915	8·5	1·458
3	2·915	8·5	1·458
4	5·099	26·0	2·550
5	1·000	1·0	0·500
	$\Sigma r^2 =$	70·0	

Rivet forces due to moment

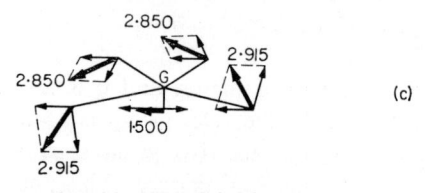

Vectorial addition of rivet forces

FIG. 91

stress multiplied by the projected area of the rivet (Fig. 92): in mild steel the bearing stress will be about 22kN/cm². It may also be necessary to consider shear failure of the member or gusset plate if a rivet hole has been placed too near the edge, or tensile failure if the rivet holes have removed too much material from the cross-section. Because of the discontinuity in structure, local stresses in joints tend to be high. For a structure subjected to many cycles of loading, fatigue failure must be considered and this may be hastened by fretting if there is relative movement between the component parts at a joint. The inaccessible interfaces at joints are often starting-points for corrosion.

Permissible shear force
= allowable shear stress
x area of cross-section

Permissible shear force
= allowable shear stress
x area of throat
[Note: area of throat = $\delta \times \frac{L}{\sqrt{2}}$
Weld sizes are usually specified by size of leg, l.]

Permissible bearing force
= allowable bearing stress
x projected area of rivet in thinner plate

The allowable shear stress in the rivet and allowable bearing stress in the plate are fictitious quantities, since the true distributions of these stresses are non-uniform

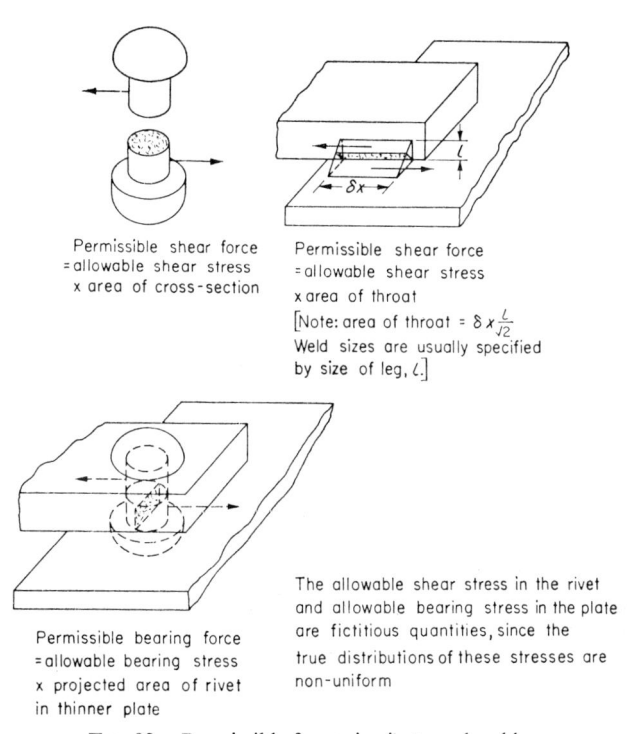

FIG. 92. Permissible forces in rivets and welds

4.5.5. *Failure of a complete structure.* In a statically determinate structure, failure of one member usually leads to failure of the whole system. In a statically indeterminate structure this is not generally so. The structure continues to support load until a sufficient number of members have failed so that the remaining system is transferred into a mechanism. We may illustrate this by considering the truss shown in Fig. 93(a). The members all have the same cross-section and there is no initial lack of fit. By the methods of Chapter 3, the elastic forces in the bars can be shown to have the values given in Fig. 93(b).

BRACED FRAMEWORKS

If we suppose that all bars yield at the same force Y in tension or compression, then as P is increased to $\frac{2}{3}Y$ bar BC will yield in tension. After bar BC has yielded, the forces may be obtained by statics: they

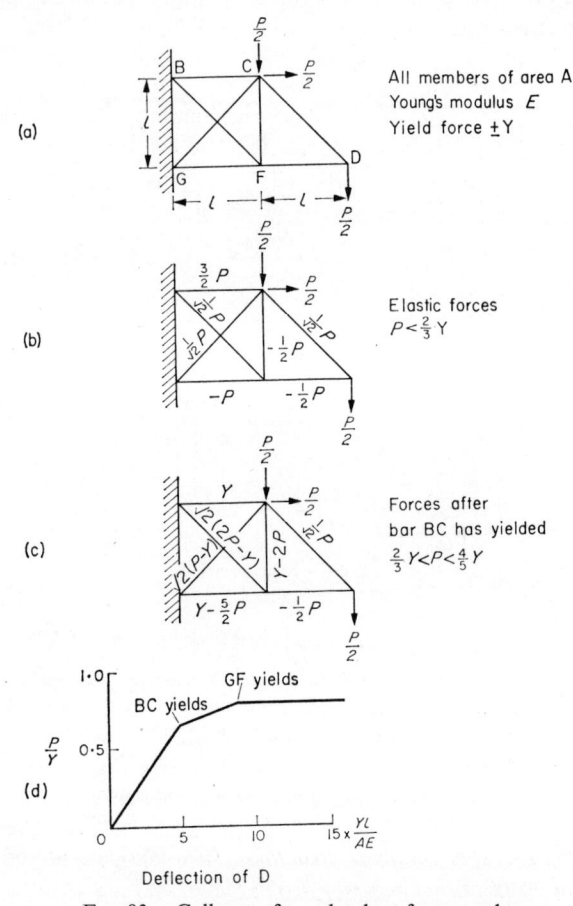

FIG. 93. Collapse of a redundant framework

are as shown in Fig. 93(c). When P is increased further to $\frac{4}{5}Y$, bar GF yields in compression and the truss collapses. The deflection of joint D is plotted in Fig. 93(d). It will be seen that after BC yields the truss becomes less stiff, but that it is still capable of carrying increased load until the second member GF yields.

SECONDARY STRESSES AND FAILURE

A redundant truss which is subjected to cyclic changes of environment may exhibit other forms of behaviour. Consider the truss shown in Fig. 94(a). All of the members have the same yield force Y in tension and compression. Members BD, BF, CD and CF have the product of Young's modulus and cross-sectional area EA: member BC has the value $\frac{1}{4}EA$. The elastic forces, which may be determined by the methods of Chapter 3, are shown in Fig. 94(b). It will be seen that if H exceeds $Y/1.019$ ($= 0.981\,Y$) member CD will yield in tension. When member CD has yielded, the forces are as shown in Fig. 94(c). The truss finally collapses due to the compressive yield of member BF when H reaches $1.2Y$.

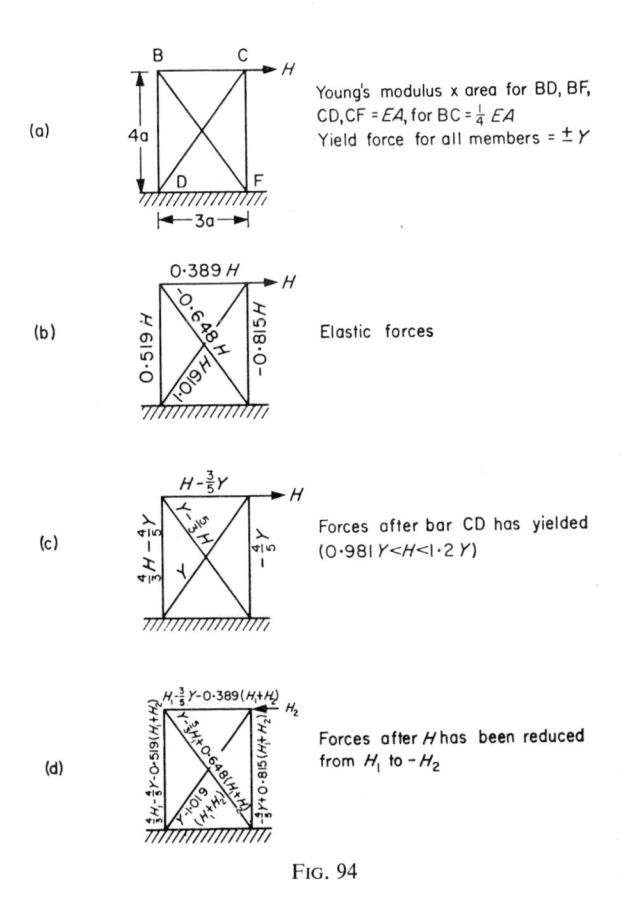

Fig. 94

155

BRACED FRAMEWORKS

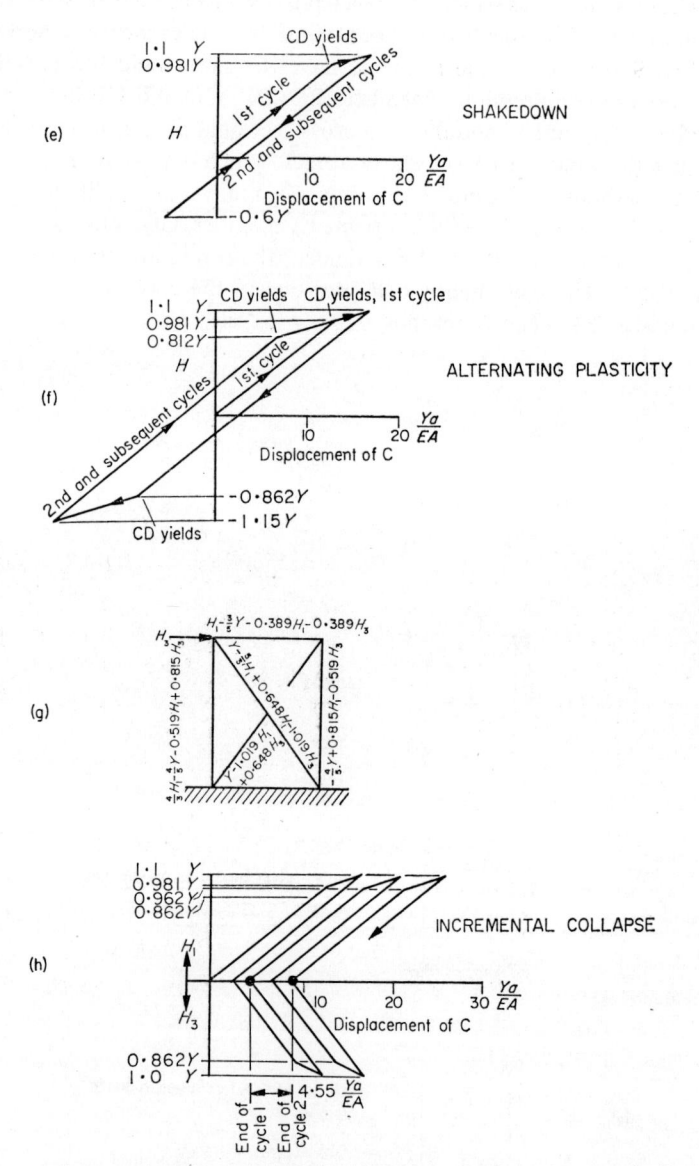

FIG. 94. Shakedown, alternating plasticity and incremental collapse

SECONDARY STRESSES AND FAILURE

Suppose that H is increased from zero to some value H_1 ($0.981Y < H_1 < 1.2Y$) and is then reduced to $-H_2$. If we assume that the changes of force as H is reduced from H_1 to $-H_2$ are elastic, the new forces in the members are as shown in Fig. 94(a). The condition that the forces shall be elastic is that bar CD shall not yield in compression, i.e.

$$Y - 1.019(H_1 + H_2) > -Y$$

or
$$H_2 < 1.962Y - H_1 \qquad (97)$$

If the applied load is now restored to H_1, the bar forces change elastically from the values shown in Fig. 94(d) to those shown in Fig. 94(c). In all subsequent cycles of loading in the range H_1 to $-H_2$ the behaviour of the truss is elastic. This type of behaviour, in which the structure exhibits inelasticity in the first cycle of loading, but in the second and subsequent cycles remains elastic, is known as "shakedown". The displacement of joint C for the particular cycle $H_1 = 1.1Y$, $H_2 = 0.6Y$ is shown in Fig. 94(e). There is a theorem due to Melan (1936) which states that if, by suitable pre-stressing, the structure could be made to support the load cycle elastically, then it will in fact shake down to an elastic state.

If the inequality (97) is not satisfied, bar CD will yield in compression. The bar forces are then similar to those shown in Fig. 93(c) but with all the signs reversed. When H is restored to H_1, bar CD again yields in tension. In each subsequent cycle of loading bar CD yields in tension and in compression, the net plastic strain being zero. This condition is known as "alternating plasticity". The displacement of joint C for the particular cycle $H_1 = 1.1Y$, $H_2 = 1.15Y$, is shown in Fig. 94(f).

Another form of behaviour is found if we consider the truss first subjected to a load H_1 ($> 0.981Y$), which is then reduced to zero, followed by a load H_3 applied to B in the same direction as H_1. The forces obtained by assuming the change from H_1 to H_3 to be elastic are shown in Fig. 94(g). It will be seen that bar BF yields in compression provided.

$$Y - \tfrac{5}{3}H_1 + 0.648H_1 - 1.019H_3 < -Y$$

or
$$H_3 > 1.962Y - H_1 \qquad (98)$$

If H_3 is now reduced to zero and H_1 is re-applied, bar CD again yields in tension. In each cycle of loading, bar CD yields in tension and bar BF yields in compression. Joints B and C slowly move in the direction BC. This type of behaviour is known as "incremental collapse". The displacement of C for the particular cycle $H_1 = 1\cdot1Y$, $H_3 = 1\cdot0Y$ is shown in Fig. 94(h). The incremental displacement per cycle is $4\cdot55\,Ya/EA$.

Shakedown of a braced framework is usually regarded as an acceptable form of behaviour. Alternating plasticity does not lead to large overall deformations, but it may produce strain fatigue if the number of cycles is large. Incremental collapse produces large deformations in a limited number of cycles and will only be acceptable in special circumstances. All of these phenomena can also occur due to temperature cycling (Parkes, 1954).

Even though the individual members of a braced framework may appear to be satisfactory, failure of a complete structure can also occur due to elastic instability. At a particular load intensity, the deflections of the structure tend to become large in a similar manner to those of the individual compression member discussed in §4.5.3. The general subject of elastic instability is outside the scope of the present book: it is discussed at length by Timoshenko and Gere (1961).

EXAMPLES

4(a). A uniform member AB, 120cm long and of flexural rigidity $0\cdot25\times10^6$ kN.cm², is held firmly at its left-hand end (A). End B is subjected to a downward deflection of 0·5cm and a clockwise rotation of 0·01 rad. Find the moments and forces applied to the member at A and B.

4(b). A uniform member CD, 144cm long and of flexural rigidity $0\cdot4\times10^6$ kN.cm², is pinned to supports at C and at D. Clockwise moments of 50kN.cm and 20kN.cm are applied at the left-hand (C) and right-hand (D) ends of the member. Support D is raised by 0·3cm. Find the rotations at C and at D.

4(c). The framework shown in Fig. 67(a) (p. 90) is subjected to a set of loads applied at B and C such that B moves 5cm to the right and 10cm down, and C moves 3cm to the right and 14cm down. If members AB and CD have flexural rigidity 2×10^4MN.m² and the remaining members 1×10^4MN.m², find the bending moments applied to the members at B by using the slope deflection equations. It may be assumed that the joints at B and C are rigid, and that the members are rigidly held at A and D.

4(d). Solve problem 4(c) by the method of moment distribution.

4(e). A uniform member of length 170cm and flexural rigidity 10^5kN.cm² is subjected to an axial compressive force of 20kN and end moments in the same sense of 25 and —15kN.cm. Determine the maximum bending moment in the member.

SECONDARY STRESSES AND FAILURE

4(f). Rewrite equation (94) as a quadratic equation in σ_m and solve this. Thence show that for large values of l/k, σ_m tends to $-\pi^2 E/(l/k)^2$, irrespective of the value of cA/Z.

4(g). A member is connected to a gusset plate by a fillet weld in the form of a U. The radius of the semicircular portion is 3cm and each straight portion is 4·25cm long. A load of 20kN is transmitted along the line of one of the straight portions of the weld. Determine the position, value and direction of the maximum force per cm run of weld.

4(h). Determine the vertical load which must be applied to joint C of the truss shown in Fig. 67(a) (p. 90) in order to cause collapse. The failing load of any member is $\pm F$.

4(i). A framework is made in the form of a square of side l, having two diagonal bracing members. All members have the same cross-sectional area A and are made from the same elasto-plastic material which has a linear stress–strain relationship of slope E up to a stress σ_y and then yields indefinitely. One diagonal member is made a proportion γ of its length too long. Equal and opposite tensile loads P are applied across the corners of the square joined by this diagonal. Determine the form of the load–deflection relationship, and show that the failing load of the framework is independent of γ.

4(j). The truss shown in Fig. 93(a) is subjected to alternating forces $P = \pm\frac{3}{4}Y$. Show that bar BC suffers alternating plasticity, and determine the magnitude of the alternating plastic strain.

5. minimum weight frameworks

5.1. Maxwell's Lemma

In the preceding four chapters of this book we have been concerned with the analysis of braced frameworks whose shape and properties were known. In this final chapter we shall consider the inverse problem—the design of a braced framework to perform certain specified functions. The practical design of a framework is usually made by repeated analysis. A guess is made as to a suitable structure and this structure is then analysed to determine whether it does in fact function satisfactorily. From the information obtained during the analysis the guess can be improved and the process repeated until a satisfactory framework is obtained. The criteria of satisfactory function, apart from the overall consideration of limitation of deflections, may include economics and aesthetics. The skill of the designer lies in making good first guesses: this skill is acquired by cumulative experience of analysis. Under certain circumstances, however, it may be possible to dispense with guesswork. For a structure which has to equilibrate safely a given set of external loads, applied in known directions at known points, and in which the criterion of good design is minimum material consumption (minimum weight) it may be possible directly to predict the perfect braced

MINIMUM WEIGHT FRAMEWORKS

framework for this purpose. We are here concerned with the direct synthesis of a perfect structure. There are two general theorems which can assist us in this process. The first, given below, is known as Clerk Maxwell's lemma (1890). We shall prove Maxwell's lemma for a plane framework, but a similar proof is valid in three dimensions.

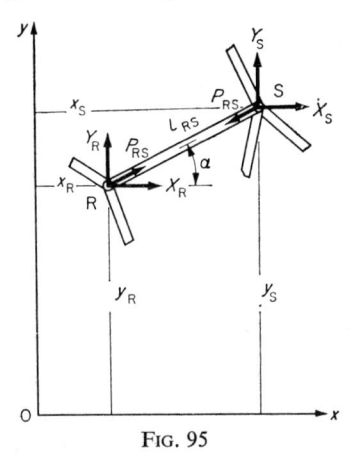

FIG. 95

Consider two joints R and S of a braced framework connected by a member RS of length l_{RS} (Fig. 95). Let the coordinates of R be (x_R, y_R) and of S (x_S, y_S) and let RS make an angle of α with Ox. The components of external load applied to joint R are X_R, Y_R and to joint S are X_S, Y_S. Then if the cross-sectional area of member RS is A_{RS} and the mean stress in it is σ_{RS}, the force in the member

$$P_{RS} = \sigma_{RS} A_{RS}$$

We may multiply both sides of this equation by l_{RS} and write

$$P_{RS} l_{RS} = \sigma_{RS} V_{RS}$$

where $V_{RS}\ (= A_{RS} l_{RS})$ is the volume of member RS. If we sum this equation for all the members connected to joint R we obtain

$$\sum_S P_{RS} l_{RS} = \sum_S \sigma_{RS} V_{RS}$$

and if we sum again for all the joints in the framework we get

$$\sum_R \sum_S P_{RS} l_{RS} = \sum_R \sum_S \sigma_{RS} V_{RS} \qquad (99)$$

The right-hand side of equation (99) is equal to twice the sum of (stress × volume) for all the bars, since each bar will be counted at

BRACED FRAMEWORKS

each of the two joints which it connects. The left-hand side of equation (99) can be expanded as

$$\sum_R \sum_S P_{RS} l_{RS} = \sum_R \sum_S \{P_{RS} \cos \alpha \cdot l_{RS} \cos \alpha + P_{RS} \sin \alpha \cdot l_{RS} \sin \alpha\}$$
$$= \sum_R \sum_S \{P_{RS} \cos \alpha (x_S - x_R) + P_{RS} \sin \alpha (y_S - y_R)\}$$
$$= \sum_R \sum_S \{P_{RS} \cos \alpha (-x_R) + P_{RS} \sin \alpha (-y_R)\}$$
$$+ \sum_R \sum_S \{P_{RS} \cos \alpha (x_S) + P_{RS} \sin \alpha (y_S)\} \quad (100)$$

Now, from equilibrium considerations,

$$\sum_S P_{RS} \cos \alpha = -X_R$$
$$\sum_S P_{RS} \sin \alpha = -Y_R$$
$$\sum_R P_{RS} \cos \alpha = X_S$$

and
$$\sum_R P_{RS} \sin \alpha = Y_S$$

If we first perform the summation in the first term on the right-hand side of equation (100) with respect to S, and if we first perform the summation of the second term with respect to R, we thus obtain

$$\sum_R \sum_S P_{RS} l_{RS} = \sum_R \{X_R x_R + Y_R y_R\} + \sum_S \{X_S x_S + Y_S y_S\}$$

Now the two terms on the right-hand side of this equation are identical, so that

$$\sum_R \sum_S P_{RS} l_{RS} = 2 \sum_R \{X_R x_R + Y_R y_R\} \quad (101)$$

Substituting from equation (99) and dividing by two,

$$\sum_{\text{bars}} (\text{stress} \times \text{volume}) = \sum_R \{X_R x_R + Y_R y_R\} \quad (102)$$

The right-hand side of equation (102) is independent of the properties of the framework: it depends only upon the values of the external loads and the positions of their points of application (the expression may readily be shown to be independent of the choice of origin 0). If we suppose that the tension members of the framework may sustain a stress σ_t and have a total volume V_t, and that the compression members may sustain a stress $-\sigma_c$ and have a total volume V_c, equation (102) becomes

$$\sigma_t V_t - \sigma_c V_c = \sum_R \{X_R x_R + Y_R y_R\} = \text{const} \quad (103)$$

MINIMUM WEIGHT FRAMEWORKS

Equation (103) shows that if all the tension members in the framework are working at stress σ_t and all the compression members at stress $-\sigma_c$, the expression $(\sigma_t V_t - \sigma_c V_c)$ is constant. The framework may be of any shape; it may be statically determinate or statically indeterminate; $(\sigma_t V_t - \sigma_c V_c)$ will always have the same value.

For minimum material consumption (minimum weight), $V_t + V_c$ must be as small as possible. From equation (103) it will be seen that if we minimise V_t, then $V_c (= \text{const} + \sigma_t V_t/\sigma_c)$ is also minimised. Similarly, if we minimise V_c, V_t is minimised. It follows that if we make either V_t or V_c as small as possible, we have a minimum weight structure. The smallest value which V_t or V_c can assume is zero. If we can devise a structure to support the applied loads in which all of the members are in tension (or all in compression), this structure will be of minimum weight. Further, all such structures which we can devise will have the same weight, whatever their configuration.

This may be illustrated by considering the loading system shown in Fig. 96(a). If we take an origin at 0, the right-hand side of equation

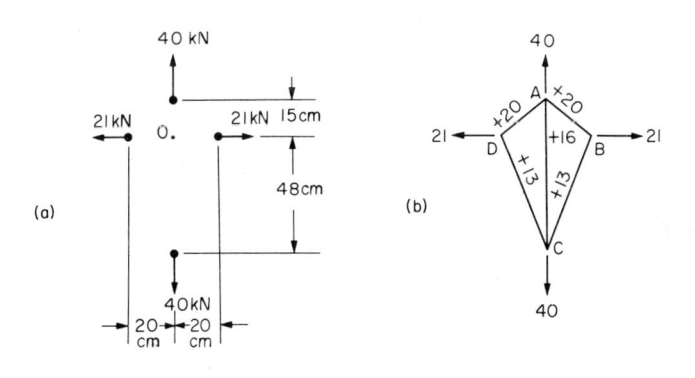

Member	Force	Area	Length	Volume
AB	+20	2·0	25	50·0
BC	+13	1·3	52	67·6
CD	+13	1·3	52	67·6
DA	+20	2·0	25	50·0
AC	+16	1·6	63	100·8
				336·0

Fig. 96

BRACED FRAMEWORKS

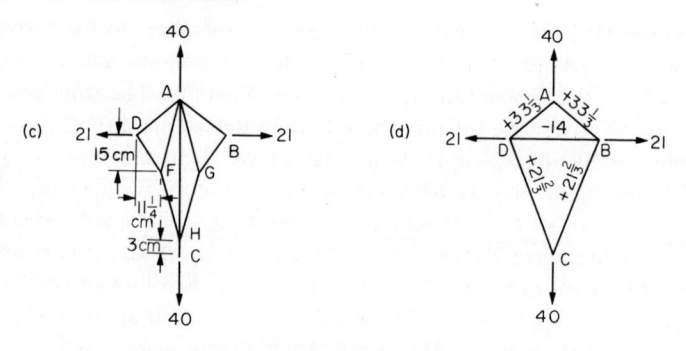

(c) and (d) diagrams of framed structures with forces shown.

(c)

Member	Force	Area	Length	Volume
AB	+16·80	1·680	25·00	42·0
BG	+12·60	1·260	18·75	23·625
GH	+18·75	1·875	31·25	58·59375
HF	+18·75	1·875	31·25	58·59375
FD	+12·60	1·260	18·75	23·625
DA	+16·80	1·680	25·00	42·0
AH	+4·00	0·400	60·00	24·0
HC	+40·00	4·000	3·00	12·0
AF	+8·25	0·825	31·25	25·78125
AG	+8·25	0·825	31·25	25·78125
				336·0

(d)

Member	Force	Area	Length	Volume
AB	+33·3˙	3·33˙	25	83·3˙
BC	+21·6˙	2·16˙	52	112·6˙
CD	+21·6˙	2·16˙	52	112·6˙
DA	+33·3˙	3·33˙	25	83·3˙
DB	−14·0	1·40	40	56·0

Volume of tension members	392·0
Volume of compression members	56·0
Sum	448·0
Difference	336·0

FIG. 96. Volumes of frameworks

(103) becomes $(21 \times 20 + 21 \times 20 + 40 \times 15 + 40 \times 48) = 3360$ kN.cm. For a limiting tensile stress of 10kN/cm², the volume of a minimum weight structure in which all the members are in tension will be $3360/10 = 336$ cm³. Such a structure is shown in Fig. 96(b). A more complicated arrangement, in which all of the bar forces are still tensile, is shown in Fig. 96(c).

The truss of Fig. 96(d) is not a minimum weight structure. Bar DB is in compression and the total volume is 448cm³ (a permissible compressive stress equal to that in tension has been assumed). The difference between the volumes of the tension and compression members is, as expected from equation (103), 336cm³.

In this example it was possible to devise structures in which all of the members were in tension. For some systems of forces, no framework can be found which equilibrates the forces and has all of its members either in tension or in compression. In order to find minimum weight structures for systems of this kind we make use of Michell's theorem.

5.2. Michell's Theorem

Suppose that we have a set of forces X_R, Y_R applied at points x_R, y_R which are to be equilibrated. Let there be a restricted domain of space D containing all the points x_R, y_R (Fig. 97). Then consider all possible frameworks S lying within D which equilibrate the given forces and which have all of their tension members at stress σ_t and

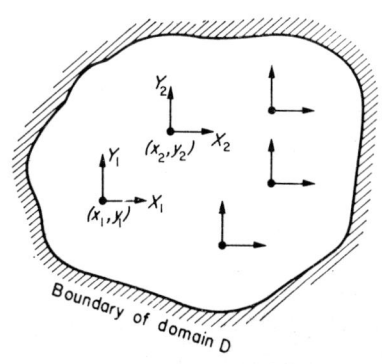

FIG. 97. Loads to be equilibrated by a framework lying within domain D

BRACED FRAMEWORKS

all of their compression members at stress $-\sigma_c$. Now suppose that the layout of one of these frameworks, S', is such that there can exist a virtual deformation of the space D containing S' so that the virtual strain along the tension members is ε' and along the compression members is $-\varepsilon'$ and that the virtual strain of any linear element of the space D does not exceed ε' in magnitude. Then Michell's theorem (1904) states that the volume V' of S' is less than or equal to the volume V of any other of the frameworks S. The proof follows.

The total volume of a framework is found by adding the volumes of the tension and compression members as

$$V = V_t + V_c$$

which can be rewritten

$$V = \frac{\sigma_t + \sigma_c}{2\sigma_t \sigma_c}(\sigma_t V_t + \sigma_c V_c) - \frac{(\sigma_t - \sigma_c)}{2\sigma_t \sigma_c}(\sigma_t V_t - \sigma_c V_c) \quad (104)$$

From equation (103) the second term in equation (104) is constant. It follows that V will be least when $(\sigma_t V_t + \sigma_c V_c)$ has its minimum value.

Consider the framework in equilibrium under the applied loads and let the virtual deformation be applied. Let the elongation of a typical member be $l\varepsilon(-\varepsilon' \leq \varepsilon \leq \varepsilon')$ and let the displacements of a typical point of load application be $u_R \varepsilon'$ and $v_R \varepsilon'$. Then the virtual work done by the system will be

$$-\sum_{\text{bars}} Pl\varepsilon + \sum_R X_R u_R \varepsilon' + \sum_R Y_R v_R \varepsilon'$$

which is zero from §2.8.3. The first term may be rewritten to give

$$-\sum_{\text{bars}} \sigma V\varepsilon + \sum_R X_R u_R \varepsilon' + \sum_R Y_R v_R \varepsilon' = 0 \quad (105)$$

For the particular case of S' where $\varepsilon = \varepsilon'$ for $\sigma = \sigma_t$ and $\varepsilon = -\varepsilon'$ for $\sigma = -\sigma_c$, we have

$$\sum_{\text{bars}} \sigma V\varepsilon = (\sigma_t V_t + \sigma_c V_c)\varepsilon'$$

and thus

$$-(\sigma_t V_t + \sigma_c V_c)\varepsilon' + \sum_R X_R u_R \varepsilon' + \sum_R Y_R v_R \varepsilon' = 0 \quad (106)$$

For any other S, ε may not have its maximum value ε' in all bars and its sign may not always coincide with that of σ. It follows that

$$\sum_{\text{bars}} \sigma V\varepsilon \leq (\sigma_t V_t + \sigma_c V_c)\varepsilon'$$

Since, from equation (105), the value of

$$\sum_{\text{bars}} \sigma V \varepsilon$$

is the same for all S, including S', we have

$$(\sigma_t V_t + \sigma_c V_c)_{S'} \leq (\sigma_t V_t + \sigma_c V_c)_S \qquad (107)$$

From equations (103), (104) and (107) it follows that the volume of framework S' is less than or equal to that of any other framework S lying within D. If D is unrestricted and includes all space, the framework S' has the absolute minimum possible volume. If D is a restricted domain, the volume of S' is smaller than or equal to that of any other framework lying within D, but it may be possible to find a framework of smaller volume than S' if that framework extends outside the assigned boundary. The volume of the optimum framework can be obtained from equations (103), (104) and (106) as

$$V' = \frac{\sigma_t + \sigma_c}{2\sigma_t \sigma_c} \left(\sum_R X_R u_R + \sum_R Y_R v_R \right) -$$
$$- \frac{(\sigma_t - \sigma_c)}{2\sigma_t \sigma_c} \sum_R \{X_R x_R + Y_R y_R\} \qquad (108)$$

It may be noted that for a system consisting only of tension members, the virtual deformation consists of a uniform strain ε' in all directions. u_R is then equal to x_R and v_R to y_R, and equation (108) reduces to equation (102). For a system consisting only of compression members, the uniform strain is $-\varepsilon'$ and thus $u_R = -x_R$ and $v_R = -y_R$ and equation (108) again reduces to (102).

The optimum framework S' has a further important property. The sum of the products of the externally applied forces and their displacements in their own lines of action is equal to twice the strain energy stored in the structure, so that for any S

$$\sum \{X_R \delta x_R + Y_R \delta y_R\} = \frac{1}{E}(V_t \sigma_t^2 + V_c \sigma_c^2)$$
$$= \frac{1}{2E} \{(V_t \sigma_t - V_c \sigma_c)(\sigma_t - \sigma_c) +$$
$$+ (V_t \sigma_t + V_c \sigma_c)(\sigma_t + \sigma_c)\} \qquad (109)$$

BRACED FRAMEWORKS

The first term on the right-hand side of equation (109) is constant for all S from equation (103). The second term is a minimum for S' from equation (107). It follows that

$$\sum \{X_R \delta x_R + Y_R \delta y_R\}$$

is smallest for framework S'. The minimum weight framework S' is thus, in a general sense, the *stiffest* of all the possible frameworks S.

5.3. The Form of the Optimum Structure

The requirement that there can exist a virtual strain of the domain of space D containing S' such that the strain along the tension members is ε', along the compression members is $-\varepsilon'$ and nowhere exceeds ε' in magnitude imposes certain limitations on the form of

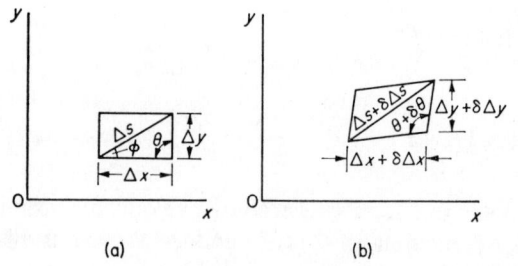

FIG. 98. Element of space. (*a*) Undeformed. (*b*) Deformed

the optimum structure. Consider an element of space having sides of length Δx and Δy as shown in Fig. 98(a). The length of the diagonal of the element, which makes an angle ϕ with the $0x$ direction, can be obtained from the cosine formula

$$\Delta s^2 = \Delta x^2 + \Delta y^2 - 2\Delta x \Delta y \cos \theta$$

where θ for the undeformed space is equal to $\frac{1}{2}\pi$. Now let the space be deformed and suppose that the element takes up the shape shown in Fig. 98(b). Then by differentiating the equation above we obtain

$$2\Delta s \cdot \delta \Delta s = 2\Delta x \cdot \delta \Delta x + 2\Delta y \cdot \delta \Delta y - 2\delta \Delta x \cdot \Delta y \cdot \cos \theta - $$
$$- 2\Delta x \cdot \delta \Delta y \cdot \cos \theta + 2\Delta x \cdot \Delta y \cdot \sin \theta \cdot \delta \theta$$

MINIMUM WEIGHT FRAMEWORKS

Dividing through by $2\Delta s^2$ and noting that $\Delta x/\Delta s = \cos\phi$, $\Delta y/\Delta s = \sin\phi$ and $\theta = \frac{1}{2}\pi$ we have

$$\frac{\delta \Delta s}{\Delta s} = \frac{\delta \Delta x}{\Delta x}\cos^2\phi + \frac{\delta \Delta y}{\Delta y}\sin^2\phi + \sin\phi\cos\phi \cdot \delta\theta$$

Now $\delta\Delta x/\Delta x$ is equal to the strain in the direction $0x$, which we may denote by ε_x, and similarly $\delta\Delta y/\Delta y = \varepsilon_y$ and $\delta\Delta s/\Delta s = \varepsilon_s$. Using these expressions for the strains and expressing $\cos^2\phi$, $\sin^2\phi$ and $\sin\phi\cos\phi$ in terms of 2ϕ, the previous equation may be written

$$\varepsilon_s = \tfrac{1}{2}(\varepsilon_x + \varepsilon_y) + \tfrac{1}{2}(\varepsilon_x - \varepsilon_y)\cos 2\phi + \tfrac{1}{2}\delta\theta \sin 2\phi \qquad (110)$$

The strain will have maximum and minimum values along those directions for which

$$\frac{\partial \varepsilon_s}{\partial \phi} = (\varepsilon_x - \varepsilon_y)\sin 2\phi + \delta\theta \cos 2\phi = 0$$

or
$$\tan 2\phi = \frac{\delta\theta}{\varepsilon_y - \varepsilon_x} \qquad (111)$$

It will be noted that the solution of equation (111) gives two values of ϕ which are $\tfrac{1}{2}\pi$ apart. Along one of these directions the strain will have its maximum value and along the other its minimum value. It follows that if we have a structure in which the tension members lie along the directions of maximum strain (ε') and the compression members along the directions of minimum strain ($-\varepsilon'$), the configuration of the structure must be such that tension and compression members always meet perpendicularly.

A further limitation on the form of the optimum structure can be found by considering the angle ϕ. Using the sine formula for a triangle we have

$$\sin\phi = \frac{\Delta y}{\Delta s}\sin\theta$$

whence on differentiating, and noting that $\theta = \tfrac{1}{2}\pi$,

$$\cos\phi \, \delta\phi = (\varepsilon_y - \varepsilon_s)\sin\phi$$

Substituting for ε_s from equation (110) we obtain

$$\delta\phi = -\tfrac{1}{2}\delta\theta + \tfrac{1}{2}\delta\theta \cos 2\phi - \tfrac{1}{2}(\varepsilon_x - \varepsilon_y)\sin 2\phi$$

BRACED FRAMEWORKS

For the particular directions where the strain is a maximum or minimum we find on substituting from equation (111) that

$$\delta\phi = -\tfrac{1}{2}\delta\theta \qquad (112)$$

Now equation (112) is true both for the direction where the strain is a maximum and for the direction where it is a minimum. It follows that lines in the directions of the maximum and minimum strains rotate by equal amounts due to the virtual deformation, and thus tension and compression members, which we have already discovered must meet perpendicularly, remain orthogonal when the space D is subjected to the virtual deformation.

Consider a framework defined by a plane net of orthogonal curvilinear coordinates α and β as shown in Fig. 99(a). The angle between the α- and x-directions at any point is ϕ. We shall refer to a line in the α-direction as "an α line": it is defined by a particular value of β. The length of an elemental chord measured in the α-direction is given by $A\delta\alpha$ and in the β direction by $B\delta\beta$, where

$$\left. \begin{array}{l} A = +\left\{\left(\dfrac{\partial x}{\partial \alpha}\right)^2 + \left(\dfrac{\partial y}{\partial \alpha}\right)^2\right\}^{\tfrac{1}{2}} \\[2mm] \text{and} \quad B = +\left\{\left(\dfrac{\partial x}{\partial \beta}\right)^2 + \left(\dfrac{\partial y}{\partial \beta}\right)^2\right\}^{\tfrac{1}{2}} \end{array} \right\} \qquad (113)$$

By drawing a line through Q parallel to PR it may be shown that

$$\frac{\partial A}{\partial \beta}.\delta\beta.\delta\alpha = -B\,\delta\beta.\frac{\partial \phi}{\partial \alpha}\delta\alpha$$

and by drawing a line through R parallel to PQ that

$$\frac{\partial B}{\partial \alpha}.\delta\alpha.\delta\beta = A\,\delta\alpha.\frac{\partial \phi}{\partial \beta}\delta\beta$$

We may simplify these relationships to

$$\frac{\partial A}{\partial \beta} = -B\frac{\partial \phi}{\partial \alpha}, \quad \frac{\partial B}{\partial \alpha} = A\frac{\partial \phi}{\partial \beta} \qquad (114)$$

Now let the framework be subjected to a virtual deformation such that the virtual displacements at any point in the α- and β-directions are u and v, and such that the strain in the α-direction is everywhere

MINIMUM WEIGHT FRAMEWORKS

ε' and in the β-direction is $-\varepsilon'$ (Fig. 99(b)). Consider the displacements of line elements PQ and PR. P moves in the α- and β-directions through u and v. Q moves through

$$u + \frac{\partial u}{\partial \alpha}.\delta\alpha \quad \text{and} \quad v + \frac{\partial v}{\partial \alpha}.\delta\alpha$$

but these displacement components are inclined at an angle $(\partial\phi/\partial\alpha)\delta\alpha$ to those of P. The component displacements of Q parallel to those of P are thus

$$u + \frac{\partial u}{\partial \alpha}.\delta\alpha - v\frac{\partial \phi}{\partial \alpha}.\delta\alpha \quad \text{and} \quad v + \frac{\partial v}{\partial \alpha}.\delta\alpha + u\frac{\partial \phi}{\partial \alpha}.\delta\alpha$$

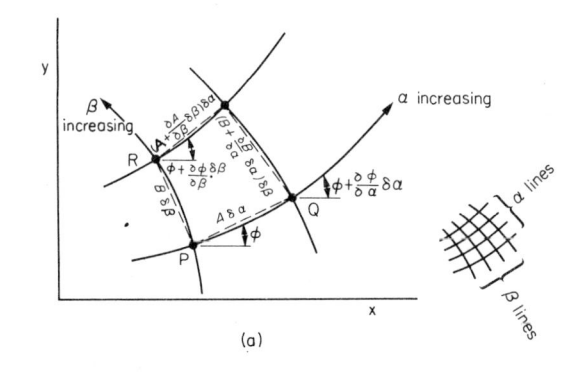

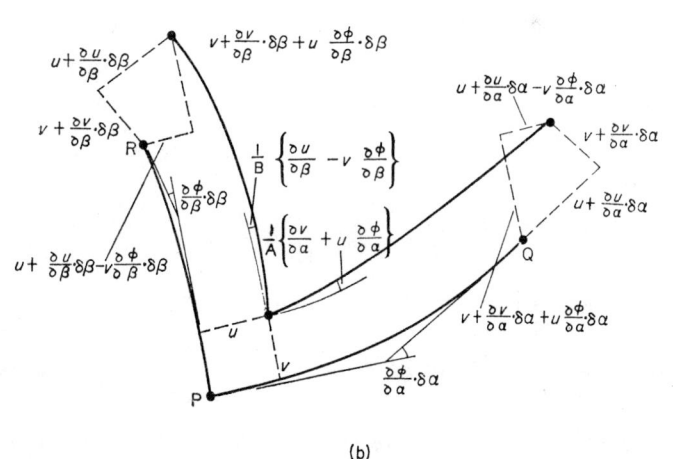

FIG. 99. Framework defined by plane net of curvilinear coordinates

BRACED FRAMEWORKS

The line element PQ extends by

$$\left(u + \frac{\partial u}{\partial \alpha} \cdot \delta\alpha - v \frac{\partial \phi}{\partial \alpha} \cdot \delta\alpha\right) - u$$

and dividing by the length $A\delta\alpha$ we obtain the linear strain in the α-direction as

$$\frac{1}{A}\left\{\frac{\partial u}{\partial \alpha} - v\frac{\partial \phi}{\partial \alpha}\right\} = \varepsilon' \qquad (115)$$

The rotation of line PQ anticlockwise is given by

$$w = \frac{1}{A\,\delta\alpha}\left\{v + \frac{\partial v}{\partial \alpha}\cdot\delta\alpha + u\frac{\partial \phi}{\partial \alpha}\cdot\delta\alpha - v\right\} = \frac{1}{A}\left\{\frac{\partial v}{\partial \alpha} + u\frac{\partial \phi}{\partial \alpha}\right\} \qquad (116)$$

R moves in the α- and β-directions through

$$u + \frac{\partial u}{\partial \beta}\cdot\delta\beta \quad \text{and} \quad v + \frac{\partial v}{\partial \beta}\cdot\delta\beta$$

but these displacement components are inclined at an angle $(\partial\phi/\partial\beta).\delta\beta$ to those of P. The component displacements of R parallel to those of P are

$$u + \frac{\partial u}{\partial \beta}\cdot\delta\beta - v\frac{\partial \phi}{\partial \beta}\cdot\delta\beta \quad \text{and} \quad v + \frac{\partial v}{\partial \beta}\cdot\delta\beta + u\frac{\partial \phi}{\partial \beta}\cdot\delta\beta$$

The line element PR extends by

$$\left(v + \frac{\partial v}{\partial \beta}\cdot\delta\beta + u\frac{\partial \phi}{\partial \beta}\cdot\delta\beta\right) - v$$

and dividing by the length $B\delta\beta$ we obtain the linear strain in the β-direction as

$$\frac{1}{B}\left\{\frac{\partial v}{\partial \beta} + u\frac{\partial \phi}{\partial \beta}\right\} = -\varepsilon' \qquad (117)$$

The rotation of line PR anticlockwise is given by

$$w = -\frac{1}{B\,\delta\beta}\left\{u + \frac{\partial u}{\partial \beta}\cdot\delta\beta - v\frac{\partial \phi}{\partial \beta}\delta\beta - u\right\} = -\frac{1}{B}\left\{\frac{\partial u}{\partial \beta} - v\frac{\partial \phi}{\partial \beta}\right\} \qquad (118)$$

Now we have already discovered that lines in the directions of the tension and compression members remain orthogonal under the

MINIMUM WEIGHT FRAMEWORKS

virtual deformation. If PQ and PR are to remain orthogonal, the rotations of PQ and PR must be equal. We have already indicated this in equations (116) and (118) by using the same symbol w. w represents the rotation of the element of the orthogonal net due to the imposition of the virtual deformation.

Multiplying equation (115) by A and differentiating with respect to β we obtain

$$\frac{\partial^2 u}{\partial \alpha \partial \beta} = v \frac{\partial^2 \phi}{\partial \alpha \partial \beta} + \frac{\partial v}{\partial \beta} \cdot \frac{\partial \phi}{\partial \alpha} + \varepsilon' \frac{\partial A}{\partial \beta}$$

Substituting for $\partial A/\partial \beta$ from (114) and for $\partial v/\partial \beta$ from (117) this becomes

$$\frac{\partial^2 u}{\partial \alpha \partial \beta} = v \frac{\partial^2 \phi}{\partial \alpha \partial \beta} - 2B\varepsilon' \frac{\partial \phi}{\partial \alpha} - u \frac{\partial \phi}{\partial \alpha} \frac{\partial \phi}{\partial \beta} \qquad (119)$$

Multiplying equation (118) by B and differentiating with respect to α we obtain

$$\frac{\partial^2 u}{\partial \alpha \partial \beta} = v \frac{\partial^2 \phi}{\partial \alpha \partial \beta} + \frac{\partial v}{\partial \alpha} \cdot \frac{\partial \phi}{\partial \beta} - B \frac{\partial w}{\partial \alpha} - w \frac{\partial B}{\partial \alpha}$$

Substituting for $\partial B/\partial \alpha$ from (114) and for $\partial v/\partial \alpha$ from (116) this becomes

$$\frac{\partial^2 u}{\partial \alpha \partial \beta} = v \frac{\partial^2 \phi}{\partial \alpha \partial \beta} - B \frac{\partial w}{\partial \alpha} - u \frac{\partial \phi}{\partial \alpha} \frac{\partial \phi}{\partial \beta} \qquad (120)$$

Subtracting (120) from (119) and dividing by B we obtain

$$\frac{\partial}{\partial \alpha}(w - 2\varepsilon' \phi) = 0$$

A similar process of elimination, using the function $\partial^2 v/\partial \alpha \partial \beta$, leads to the equation

$$\frac{\partial}{\partial \beta}(w + 2\varepsilon' \phi) = 0$$

These two equations are similar to those given by Prager and Hodge (1951) for the shear lines in perfectly plastic solids. The rotation w may be eliminated between them to give

$$\frac{\partial^2 \phi}{\partial \alpha \partial \beta} = 0 \qquad (121)$$

BRACED FRAMEWORKS

Equation (121) imposes a restriction upon the possible forms of Michell frameworks. A direct geometrical interpretation of equation (121) may be obtained if we integrate with respect to each variable in turn. Starting with α we have

$$\frac{\partial \phi}{\partial \beta} = f(\beta)$$

and then integrating with respect to β we find

$$\{\phi_2 - \phi_1\}_{\alpha \, \text{const}} = F(\beta_1, \beta_2)$$

The geometrical meaning of this equation is shown in Fig. 100(a) where the sides of the curvilinear rectangle may be finite. *The angle formed by the tangents of two α-lines at their points of intersection with*

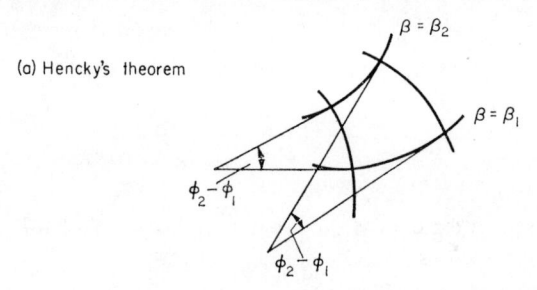

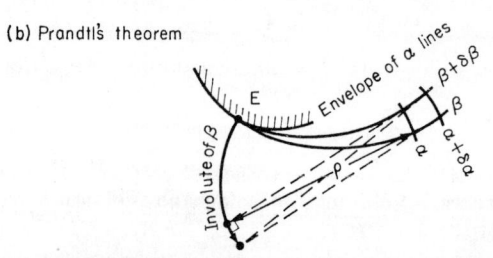

Fig. 100. Hencky's theorem and Prandtl's theorem

a single β-line does not depend upon the choice of that β-line. (Similar reasoning applies if we exchange α and β.) This statement is equivalent to Hencky's theorem (1923) for the form of the shear lines in perfectly plastic solids: it will be known by the same name here. A useful corollary of Hencky's theorem may be noted: *if one α-line is straight, all the α-lines are straight.*

A second geometrical interpretation of equation (121) may be obtained by considering the radius of curvature ρ of a β-line (say PR in Fig. 99(a)). We have

$$\rho = \frac{B \, \partial \beta}{(\partial \phi / \partial \beta) \cdot \delta \beta} = B \left(\frac{\partial \phi}{\partial \beta} \right)^{-1}$$

The variation of ρ with α is obtained by differentiation as

$$\frac{\partial \rho}{\partial \alpha} = \frac{\partial B}{\partial \alpha} \left(\frac{\partial \phi}{\partial \beta} \right)^{-1} - B \left(\frac{\partial \phi}{\partial \beta} \right)^{-2} \frac{\partial^2 \phi}{\partial \beta \, \partial \alpha} = A$$

on substituting from equations (114) and (121). We thus find that on moving from P to Q (Fig. 99(a)) the radius of curvature of the β-lines changes by $A \, \delta \alpha$, which is equal to the distance PQ. The centre of curvature of the β-lines thus moves perpendicular to the tangent to the α-line, as shown in Fig. 100(b). It follows that *as we move along a given α-line, the locus of the centres of curvature of the β-lines forms an involute of the given α-line* (and similarly if we exchange α and β). This is equivalent to Prandtl's theorem (1923) in plasticity and it will be known by that name here, although Michell (1904) has a prior claim. Orthogonal nets satisfying equation (121) are often known as Hencky–Prandtl nets.

At the point E (Fig. 100(b)), where the involute intersects the α-line, the radius of curvature of the β-lines is zero and adjacent α-lines intersect. It follows that E lies on the envelope of the α-lines and that the β-lines are cusped on this envelope and thus cannot cross it. *The envelope of the α-lines forms a boundary to the system of β-lines (and vice versa)*.

5.4. Michell Frameworks

There are many forms of orthogonal net satisfying equation (121). The most useful of these are the ones which extend to the whole of space, since frameworks formed of these nets will be of absolutely minimum volume. The simplest net of this type is the rectangular net (Fig. 101(a)). Equiangular spirals (Fig. 101(b)) are also unrestricted except for their origins. Circular fans (Fig. 101(c)) can extend to most of space, but it is important to note that the fans cannot be closed into complete circles, since there is a strain incompatibility at the radius of closure. Nets of different kinds may be combined provided the virtual strains are compatible at the boundaries (it is necessary that the strain in the direction of the boundary and the rotation should agree—the strain perpendicular to the boundary may be discontinuous).

(a) Rectangular net

(b) Equiangular spirals

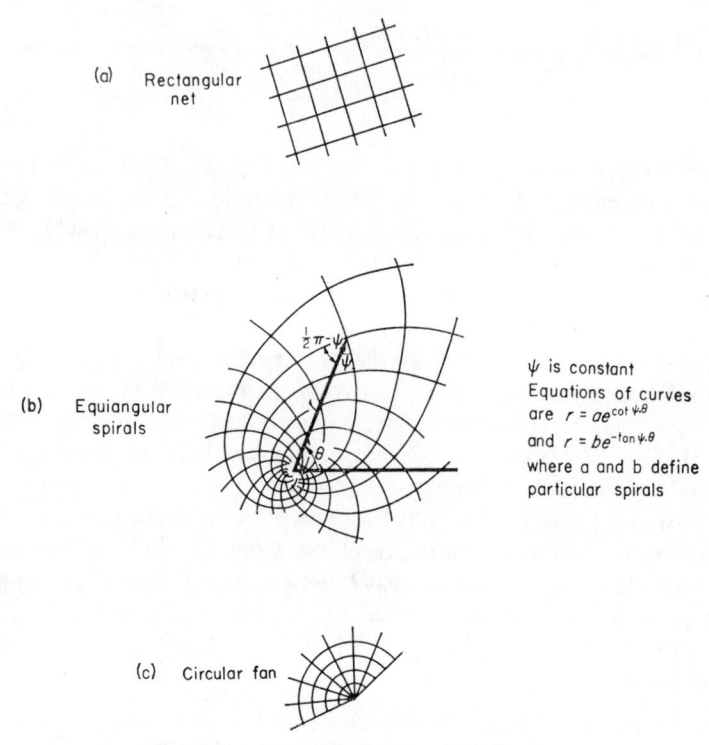

ψ is constant
Equations of curves
are $r = ae^{\cot\psi\cdot\theta}$
and $r = be^{-\tan\psi\cdot\theta}$
where a and b define particular spirals

(c) Circular fan

Fig. 101. Nets extending to most of space

When we endeavour to make frameworks from these nets we have to consider how a concentrated load may be introduced into the system. This may be done by introducing especially heavy members, or by ensuring that a point in the framework where many members intersect coincides with the load. This is illustrated by our first example, that of the single vertical force supported by two equidistant vertical reactions applied at the same height. A suitable net and the corresponding Michell framework is shown in Fig. 102(a). The radii

MINIMUM WEIGHT FRAMEWORKS

of fan A have virtual strain ε', and the radii of fan B have virtual strain $-\varepsilon'$. These strains are compatible with those in the rectangular nets C and D at their boundaries with the fans. The decrease in angle of fan A due to the virtual strain system is equal to the increase in

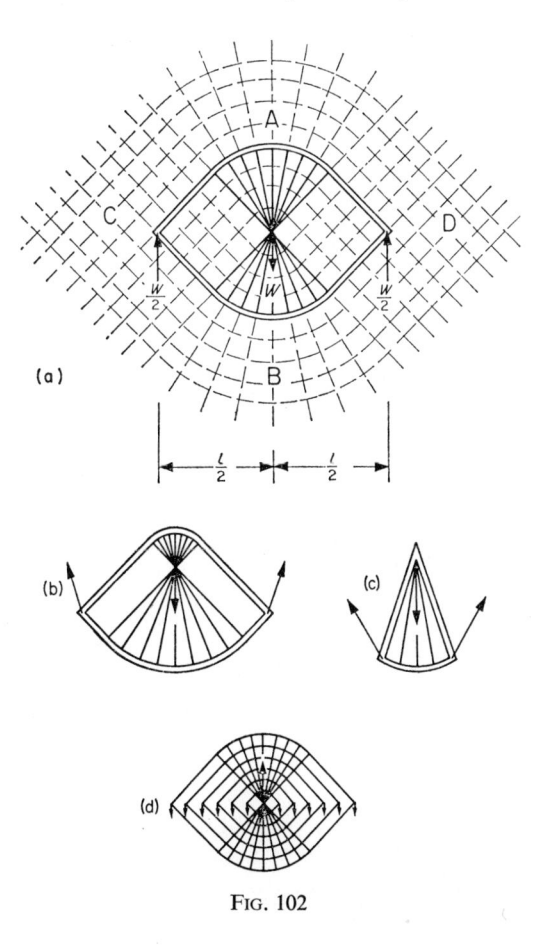

Fig. 102

angle of fan B, and thus the rectangular nets remain orthogonal. The origin of the fans is a singular point apparently disobeying our requirement that tension and compression members must meet perpendicularly. The virtual strain system here is, however, a compatible one. The argument used in §5.3 depends on the existence of

BRACED FRAMEWORKS

a very small element of distorted space and it cannot be expected to apply to a point of zero dimensions.

Since the net shown in Fig. 102(a) extends to all space, the Michell framework is of absolute minimum volume. This volume can be obtained from equation (108) if we note that due to the virtual strain system the points of application of the reactions rise by $(\frac{1}{2}+\frac{1}{4}\pi)l\varepsilon'$. We then have

$$V' = \frac{\sigma_t+\sigma_c}{2\sigma_t\sigma_c}\left\{2\frac{W}{2}(\tfrac{1}{2}+\tfrac{1}{4}\pi)l\right\}$$

$$= \frac{\sigma_t+\sigma_c}{8\sigma_t\sigma_c}(2+\pi)Wl \tag{122}$$

The net shown in Fig. 102(a) can be used to find a Michell framework for other load systems. Examples are shown in Fig. 102(b)–(d).

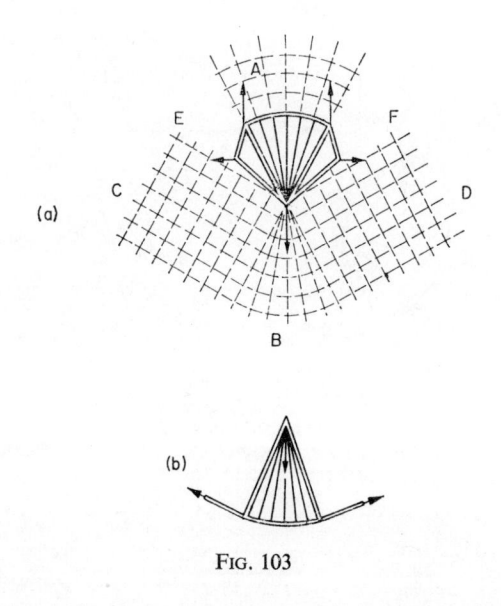

FIG. 103

It is not necessary that all of the component nets in our system should have equal and opposite maximum virtual strains ε' and $-\varepsilon'$. A uniform dilatation of space, with the same strain (either ε' or $-\varepsilon'$) in all directions, may also be introduced. In such a region, the framework will be of the simple type discussed in §5.1: it may have any configuration, and the forces in all the members in this region will have the same sign. As an example we may consider the net of Fig. 103(a), and the corresponding Michell framework used to equilibrate the five-load system shown. Regions A, B, C and D have similar characteristics to those in Fig. 102(a). In regions E and F the virtual

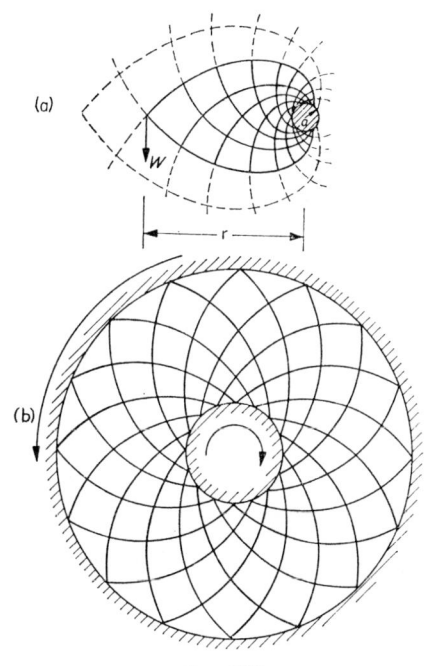

FIG. 104

strain is uniform dilatation in all directions of magnitude ε'. The net of Fig. 103(a) can be used to find Michell frameworks for other load systems: an example is shown in Fig. 103(b).

In Fig. 104(a) we show the equiangular spiral net with $\psi = \tfrac{1}{4}\pi$ used to solve the problem of transmitting a force to a circular shaft. The net satisfies the condition that the virtual strain tangential to the surface of the shaft shall be zero. The absolute minimum volume of the framework is

$$V' = \frac{\sigma_t + \sigma_c}{\sigma_t \sigma_c} Wr \ln \frac{r}{a}$$

This net can also be used for the problem of transmitting a torque between two coaxial circles, as shown in Fig. 104(b).

An interesting three-dimensional solution given by Michell (1904) is that for the minimum-weight shaft. This is shown in Fig. 105. All the rhumb lines of one system have virtual strain ε' and all those of

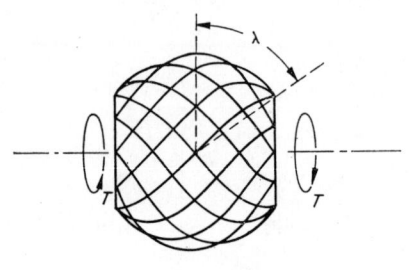

Fig. 105

the other system have virtual strain $-\varepsilon'$. The third system of orthogonals, the radii from the centre of the sphere, have zero virtual strain and simply rotate about the axis of torque. The minimum volume is

$$V' = \frac{2(\sigma_t + \sigma_c)}{\sigma_t \sigma_c} T \ln \tan(\tfrac{1}{4}\pi + \tfrac{1}{2}\lambda)$$

All of the solutions which we have considered so far have used unrestricted nets extending to the whole of space: the frameworks were thus of absolute minimum volume. If we require our framework to be restricted to a domain D, Michell's theorem still enables us to find the framework of minimum volume within D, but this volume may be greater than the absolute minimum volume obtained when the domain is unrestricted. As an example we consider the problem of Fig. 102(a) but with the restriction that the structure must lie entirely above the line joining the points of application of the forces. A suitable net is shown in Fig. 106. It is left to the reader to show that this net cannot be extended to the whole of space. Noting that the points of application of the reactions rise by $\tfrac{1}{2}\pi l \varepsilon'$ due to the virtual strain system, we find from equation (108) that the minimum volume is

$$V' = \frac{\sigma_t + \sigma_c}{4\sigma_t \sigma_c} \pi W l$$

FIG. 106

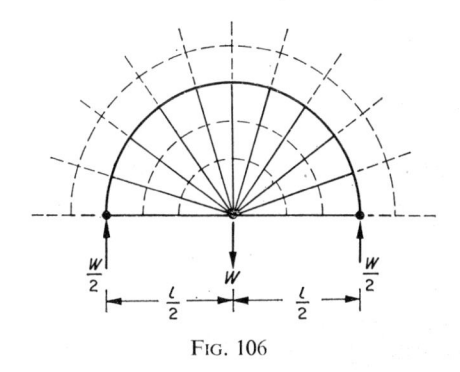

This volume is $2\pi/(2+\pi) = 1\cdot22$ times that of the frame of absolute minimum volume. Our requirement that the frame shall lie on one side only of the line joining the points of application of the forces thus involves a weight penalty of 22 per cent.

5.5. The Symmetrical Three-force System

The nets shown in Figs. 102 and 103 enable us to solve a general system of the type shown at the top of Fig. 107. The analysis is formally limited to $0 \leq \theta \leq \pi$ and it follows that for equilibrium $-\frac{1}{2}\pi < \phi - \theta < \frac{1}{2}\pi$. We have to distinguish five cases, and it will be sufficient if we investigate them for the range $0 \leq \theta \leq \frac{1}{2}\pi$, since behaviour in the range $\frac{1}{2}\pi \leq \theta \leq \pi$ can be derived by rotation about both $\theta = \frac{1}{2}\pi$ and $\phi = \frac{1}{2}\pi$. The permissible stress σ is the same in tension and in compression.

For $\phi \leq 0$, the framework has forces of the same sign in all members. The volume can be found from equation (102) as

$$V' = \frac{Wr}{\sigma} \sec(\phi - \theta) \cos \phi$$

For $0 < \phi \leq \frac{1}{2}\pi$, $\theta \leq \frac{1}{4}\pi$, the framework is of the type shown in Fig. 102(c) and the volume is

$$V' = \frac{Wr}{\sigma} \sec(\phi - \theta) \{\cos \phi + 2\theta \sin \phi\}$$

BRACED FRAMEWORKS

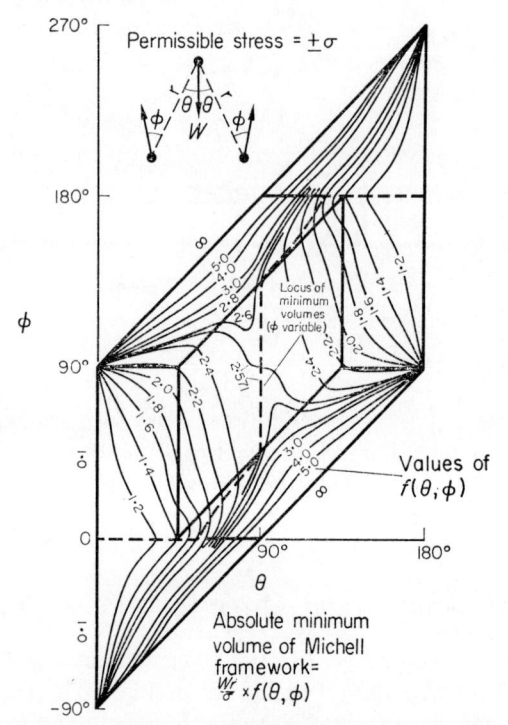

FIG. 107. Absolute minimum volumes of frameworks

For $-\tfrac{1}{4}\pi < \phi-\theta < \tfrac{1}{4}\pi$, $\theta > \tfrac{1}{4}\pi$, the framework is of the type shown in Fig. 102(b) and the volume is

$$V' = \frac{Wr}{\sigma}\sec(\phi-\theta)\{\sin(2\theta-\phi)+\tfrac{1}{2}\pi\sin\phi\}$$

For $\phi-\theta \geqq \tfrac{1}{4}\pi$, $\phi > \tfrac{1}{2}\pi$, the framework is of the type shown in Fig. 103(b) and the volume is

$$V' = \frac{Wr}{\sigma}\sec(\phi-\theta)\{-\cos\phi+2(\theta-\phi+\tfrac{1}{2}\pi)\sin\phi\}$$

For $\phi-\theta \leqq -\tfrac{1}{4}\pi$, $\phi > 0$, the framework is similar to the previous type, but inverted. The volume is

$$V' = \frac{Wr}{\sigma}\sec(\phi-\theta)\{\cos\phi+2(\phi-\theta+\tfrac{1}{2}\pi)\sin\phi\}$$

MINIMUM WEIGHT FRAMEWORKS

These volumes, which are all absolute minimum values, may be written in the form $V' = (Wr/\sigma)f(\theta,\phi)$. In Fig. 107, $f(\theta,\phi)$ is plotted for the full range of values of θ and ϕ (the heavy lines define the boundaries of the five modes of behaviour given above).

It is of interest to consider the problem of deciding upon the best direction for the reactions, when choice is available. For $\theta \leq 45°$, the minimum volume clearly occurs for $\phi = 0$. For $\theta > 45°$, however, the minimum volume occurs somewhere in the zone of Fig. 107 for which $\phi - \theta \leq -\frac{1}{4}\pi$ and $\phi > 0$. In this zone

$$V' = \frac{Wr}{\sigma}\sec(\phi-\theta)\{\cos\phi + 2(\phi-\theta+\tfrac{1}{2}\pi)\sin\phi\}$$

Differentiating with respect to ϕ, we find that V' is a minimum when

$$2(\phi-\theta+\tfrac{1}{2}\pi)\cos\theta + \sin(2\phi-\theta) = 0 \qquad (123)$$

When $\phi = 0$, equation (123) gives $\theta = 52\cdot6°$. For $\theta \leq 52\cdot6°$ the minimum volume thus still occurs for $\phi = 0$. For $\theta > 52\cdot6°$, equation (123) has solutions as shown in the table below.

$\theta =$	52·6°	60°	70°	80°	90°
$\phi =$	0·0°	8·7°	20·5°	32·5°	45·0°

The locus of minimum volumes when ϕ can be varied is shown by the dotted line in Fig. 107: it will be noted that for $\theta = 90°$, ϕ can have any value between 45° and 135°. The minimum volume is shown as a function of θ in Fig. 108.

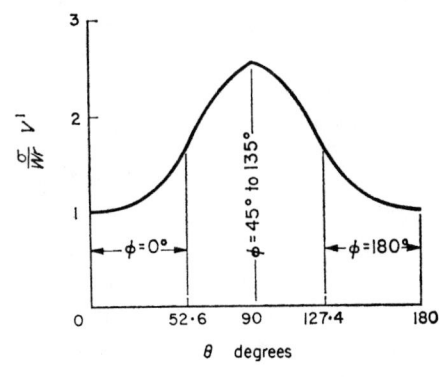

FIG. 108. Minimum volume when ϕ may be varied

Another problem of interest is that of supporting a central load between abutments a fixed distance l apart. In this case $r = \frac{1}{2}l\operatorname{cosec}\theta$, so that if we multiply the ordinates of Fig. 108 by $\frac{1}{2}\operatorname{cosec}\theta$, we obtain the function $(\sigma/Wl)V'$. This has a minimum at $\theta = 45°$ when $\phi = 0°$, so that the most economical way of supporting a central load between abutments is by means of straight bars having slopes of 45°.

5.6. Graphical Methods

Analytical solutions are known for only a small number of minimum volume frameworks. It is sometimes possible, however, to obtain solutions graphically. Graphical methods have been well developed for problems in plasticity, and various types of solution are discussed by Prager and Hodge (1951) and Johnson and Mellor (1962). We shall content ourselves here with one method, which can be used when two intersecting lines of the orthogonal net are given. Let these lines be as shown in Fig. 109(a). Divide the given curves, from their point of intersection, into segments for which the change of angle $\Delta\phi$ is constant (Fig. 109(b)). Denote the points obtained by their coordinates (α, β) as shown. Then, if we draw in the elemental chords, these will all meet at angle $\Delta\phi$ (Fig. 109(c)). We next erect from each of the points a line perpendicular to the *following* elemental chord. The intersection of the lines from $(1,0)$ and $(0,1)$ defines the point $(1,1)$. From $(1,1)$ we draw lines perpendicular to those from $(2,0)$ and $(0,2)$ and thus obtain the points $(2,1)$ and $(1,2)$. By a similar process we can obtain the points $(3,1)$, $(4,1)$, $(5,1)$, etc., and the points $(1,3)$, $(1,4)$, $(1,5)$, etc. The two lines defined by these points can now be used in exactly the same way as the original pair of lines to obtain further points, and the net can thus be completed. Once the points on the original lines have been defined, the process of obtaining the net requires only the use of a 90° square. For accuracy, $\Delta\phi$ should be kept fairly small: a value of 15° is often used.

That the net obtained satisfies Hencky's theorem may be seen if it is noted that at each intersection each line turns through $\Delta\phi$. All intersections of the net have similar geometry—a fact which is very useful when resolving at the joints to find the forces in the bars. If the original lines curve in opposite directions, a similar process may be used, but in order to maintain similar angles for the elementary

MINIMUM WEIGHT FRAMEWORKS

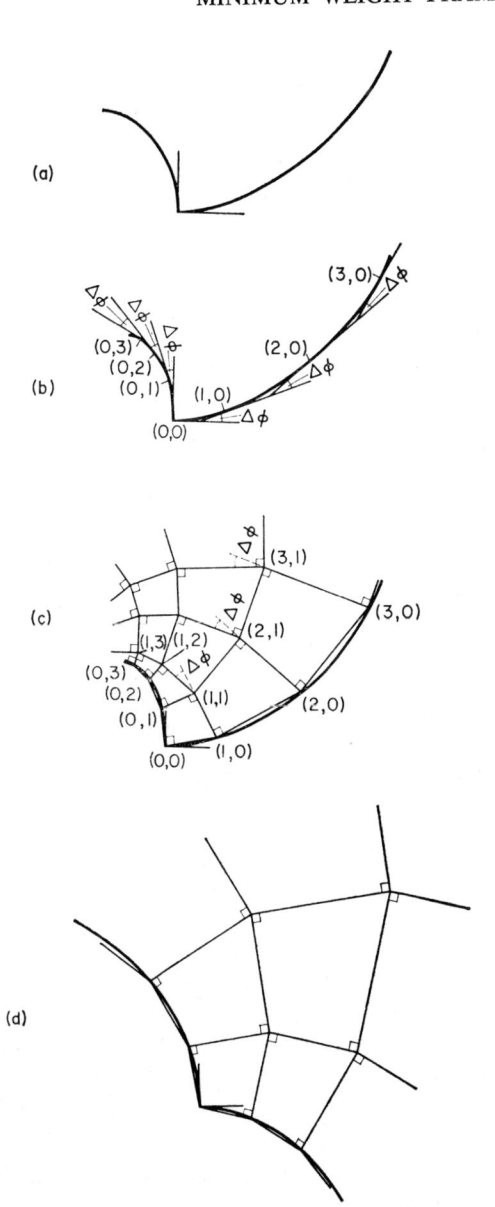

Fig. 109. Graphical construction of net

BRACED FRAMEWORKS

rectangles, the lines must be drawn perpendicular to the *preceding* elemental chords. This is shown in Fig. 109(d).

We shall use the graphical method to extend the orthogonal net formed by two equal circular fans and a square, shown in Fig. 110(a). There is a strain incompatibility between two sides of the square and

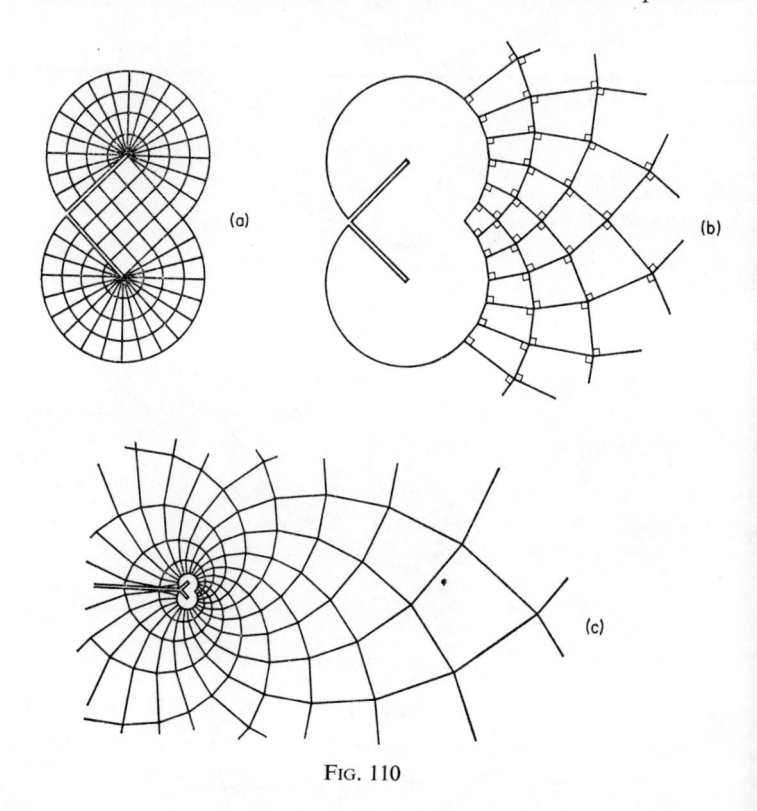

Fig. 110

the circles, indicated by the double lines. Using the method shown in Fig. 109(d), we can extend the net from the outer edges of the fans, as shown in Fig. 110(b). The complete net is shown in Fig. 110(c): it extends to the whole of space with the exception of the double lines. This net is convenient for the solution of certain three-load problems. As an example we may consider the load system shown in Fig. 111(a). We cannot with the present net specify the value of V initially. A suitable net with $\Delta\phi = 15°$ is shown in Fig. 111(b). The framework is statically determinate and the forces in the members can be found by resolution at the joints, starting at the 5kN load. When these forces have been determined, the vertical reactions can be obtained,

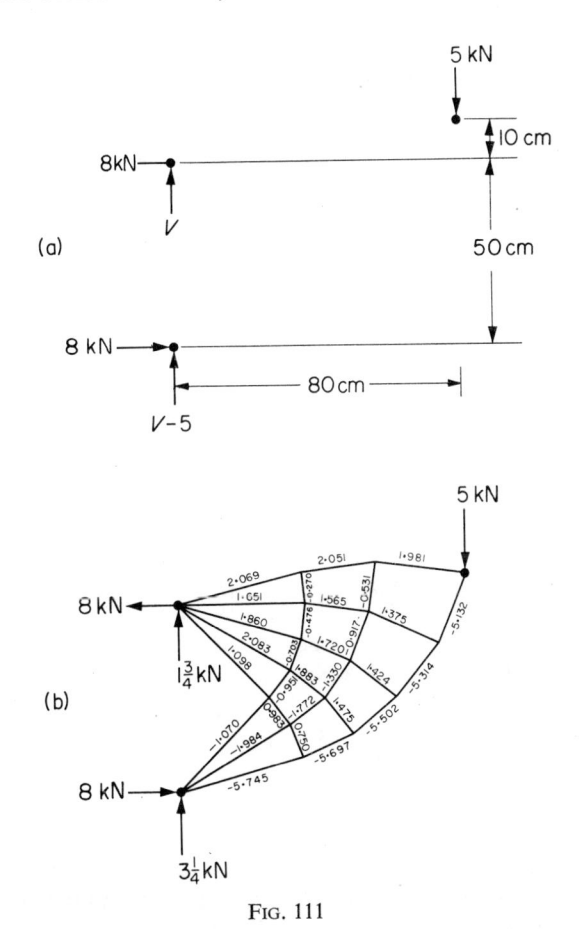

FIG. 111

as shown in Fig. 111(b). For this particular load system the framework shown is (approximately) a Michell structure. Its volume can be found by summation over the bars of (force × length/permissible stress). For a permissible stress of 10kN/cm², the volume of the tension members is 56cm³ and of the compression members is 77cm³, making a total of 132cm³. The difference in the volumes of tension

187

BRACED FRAMEWORKS

and compression members, 21cm³, can be checked by equation (103). It is of interest to compare this volume with that of a three-bar structure connecting the points of load application: the volume of this structure is 179cm³.

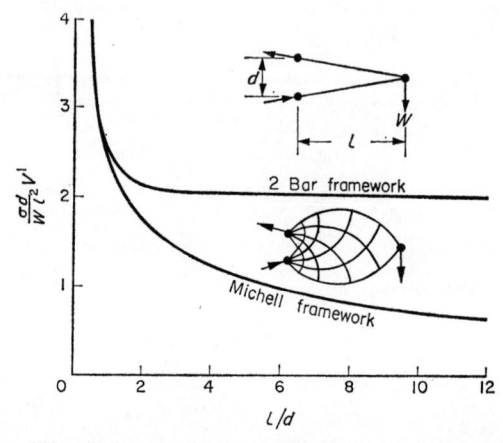

Fig. 112. Volumes of symmetrical cantilevers (after Chan)

If the vertical load is applied at a height midway between those of the two reactive forces, the net is symmetrical and the reactive forces intersect at the point of load application. Chan (1960) has investigated the symmetrical cantilever partly by graphical means and partly analytically. He obtains the relationship between volume and aspect ratio of the structure shown in Fig. 112. The corresponding relation for the two-bar structure is shown for comparison. Chan has also applied a combination of two nets of this type to the four-load (pure bending) problem.

All of the structures discussed in the present section have been derived from the net shown in Fig. 110(c) which has a limited domain. They are therefore structures of minimum volume for that domain, but it is possible that they are not of absolute minimum volume.

Other Michell structures, and more general aspects of minimum weight design, are discussed by Hemp (1973).

MINIMUM WEIGHT FRAMEWORKS

EXAMPLES

5(a). The symmetrical redundant framework shown in Fig. 113 has all of its members sustaining stresses of $\pm 10 \text{kN/cm}^2$. Show by direct calculation that provided the force R in the member shown lies between 0 and 30kN, the total volume of the framework is independent of R and equal to $300\sqrt{3}\text{cm}^3$.

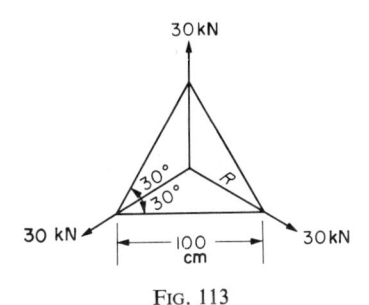

FIG. 113

Confirm this result by using equation (103). Determine the volume of the framework if R exceeds 30kN.

5(b). Determine the volume of material in the tension members and in the compression members of the truss shown in Fig. 25 (p. 36), assuming that the permissible stress in tension is 9kN/cm^2 and in compression is 7kN/cm^2. Show that your answers are consistent with equation (103).

5(c). Show that all of the members of the framework shown in Fig. 114 are in tension, and find its volume if the permissible stress is 200MN/m^2.

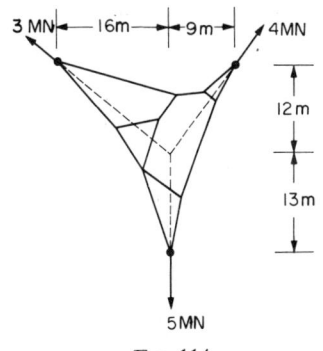

FIG. 114

5(d). Show that if a virtual strain field in the form of a circular fan of semiangle θ and radius r is placed with its axis of symmetry vertical, then the vertical and horizontal virtual displacements of the tips of the fan are $(\cos\theta + 2\theta\sin\theta)r\varepsilon'$ and $(\sin\theta - 2\theta\cos\theta)r\varepsilon'$ respectively.

189

BRACED FRAMEWORKS

5(e). Use the virtual strain field of Fig. 102(a) (p. 177) to derive a Michell structure for the load system shown in Fig. 115. Find the vertical and horizontal virtual deflections of the points of load application and determine the volume of the structure from equation (108). The permissible stress in tension is 10kN/cm^2 and in compression is 8kN/cm^2. Check your answer by direct analysis of the forces in the members.

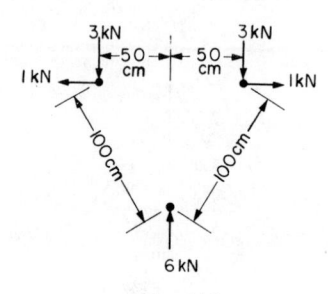

Fig. 115

answers to examples

1(a). No. The doors will jam if the depth of snow exceeds 0·47m.

1(c). (i) and (iv).

1(d). 1·0010m.

1(e). 18·3MN/m^2.

1(f). (i) $2·45 \times 10^{-3}$; (ii) $1·50 \times 10^{-3}$; (ii) $1·20 \times 10^{-3}$.

1(g). 0·00499, 0·0488, 0·405.

1(h). 2·0214m.

1(j). (i) and (iv).

1(k). (i) no significant change; (ii) considerable increase; (iii) truss collapses (mechanism).

2(a) and (b). $P_{CD} = -50$, $P_{BC} = 10\sqrt{10}$, $P_{FD} = -30\sqrt{2}$, $P_{FG} = 5\sqrt{17}$, $P_{BF} = -25\sqrt{5}$, $P_{BG} = 35\sqrt{2}$, $P_{AB} = -55$kN.

2(c). 7·11kN.

2(d). $P_{AB} = 12\sqrt{2}$, $P_{BC} = 0·8$, $P_{CD} = 23·2$, $P_{DF} = 12\sqrt{2}$, $P_{FG} = 12$, $P_{GH} = 62·4$, $P_{HJ} = 40$, $P_{JA} = 12$, $P_{AF} = -12$, $P_{BH} = -\frac{16}{5}\sqrt{74}$, $P_{DH} = -\frac{32}{5}\sqrt{74}$, $P_{CJ} = -5\sqrt{89}$, $P_{CG} = -\frac{39}{5}\sqrt{89}$kN.

2(e). $P_{DA} = P_{DB} = \frac{10}{3}\sqrt{5}$, $P_{AF} = P_{BG} = \frac{10}{3}\sqrt{5}$, $P_{FC} = P_{GC} = -\frac{20}{3}\sqrt{5}$, $P_{DF} = P_{DG} = -\frac{10}{3}\sqrt{3}$, $P_{FG} = \frac{20}{3}\sqrt{3}$, $P_{DH} = \frac{20}{3}\sqrt{5}$, $P_{HF} = P_{HG} = -\frac{10}{3}\sqrt{5}$kN.

2(f). (ii) and (iii).

2(g). $-\frac{54500}{224}\sqrt{65} = -1960$kN (occurs when 500kN load is 19m from left-hand support).

2(h), (i) and (j). 1·463cm upwards.

2(k). Down and to the right, 10^{-3}cm, A (36, 76), B (179, 44), C (238, 12), D (0, 12), G (179, 0), H (238, 64), J (0, 128).

ANSWERS TO EXAMPLES

2(l). (i) 6·8cm; (ii) 0; (iii) 7·3cm. Hooke's law is not valid.

2(m). $\dfrac{Pl}{6AE}(41+14\sqrt{2})$.

2(n). $\dfrac{1}{2\sqrt{3}} \cdot \dfrac{W}{AE}$ m.

3(a). 27·7MN.

3(b). 10·2kN.

3(c). $Pl^2kr^3 = \dfrac{v}{r}\left(1-\dfrac{v}{r}\right)\left(2-\dfrac{v}{r}\right)$

3(d). —14·8, —6·6kN.

3(e). 1·54, 3·39kN.

3(f). —26·7kN.

3(g). —1·74, 1·11kN.

3(h). $-\dfrac{P}{4n-3}\left\{1+2\cos\dfrac{(r-1)}{2n}\pi\right\}$.

3(i). 0·033cm down.

3(j). $0{\cdot}747(-0{\cdot}353)^{n-2}P$.

3(k). —273N.

4(a). 10·4kN.cm anticlockwise and 0·174kN down at A; 51·3kN.cm clockwise and 0·174kN up at B.

4(b). 0·00272 rad clockwise, 0·00268 rad anticlockwise.

4(c) and (d). —4·3, 0·4, 3·8MN.m applied to BA, BD, BC.

4(e). 55·5kN.cm.

4(g). At tip of U; 2·30kN/cm at $31\tfrac{1}{2}°$ to direction of load.

4(h). 1·2F.

4(i). Straight lines joining points:

$P = 0$ $\qquad\qquad\qquad\qquad$ $\delta = 0$

$P = \dfrac{(2+2\sqrt{2})\sigma_y A+\sqrt{2}\,\gamma AE}{(2+\sqrt{2})}$ $\qquad$ $\delta = \sqrt{2}l\left\{\dfrac{\sigma_y}{E}+\dfrac{\sqrt{2}\gamma}{(2+2\sqrt{2})}\right\}$

$P = 2\sigma_y A$ $\qquad\qquad\qquad\qquad$ $\delta = \sqrt{2}l\left\{(1+\sqrt{2})\dfrac{\sigma_y}{E}-\dfrac{(2+\sqrt{2})\gamma}{(2+2\sqrt{2})}\right\}$

$P = 2\sigma_y A$ $\qquad\qquad\qquad\qquad$ $\delta = \infty$

4(j). $(2{\cdot}5+\sqrt{2})\dfrac{Y}{AE}$

5(a). $100\sqrt{3}(2R-30)\text{cm}^3$.

5(b). 452, 638cm³.

5(c). 0·925m³.

5(e). $100\left(\dfrac{\sqrt{3}}{2}+\dfrac{\pi}{6}\right)\varepsilon'$, $100\left(\dfrac{\pi\sqrt{3}}{6}-\dfrac{1}{2}\right)\varepsilon'$, $\tfrac{5}{4}\{10(3\sqrt{3}-1)+3(\sqrt{3}+3)\pi\}$ cm³.

references

ASPLUND, S. O. (1966). *Structural Mechanics: Classical and Matrix Methods.* Prentice-Hall.
BAKER, J. F. and OCKLESTON, A. J. (1935). Aer. Res. Cttee., R. and M. 1667.
BAKER, J. F. HORNE, M. R. and HEYMAN, J. (1956). *The Steel Skeleton*, vol. 2, C.U.P.
BEGGS, G. E. (1927). *J. Franklin Inst.* **203**, 375.
BENDIXEN, A. (1914). *Die Methode der Alpha-Gleichungen zur Berechnung von Rahmenkonstruktionen*, Berlin.
BETTI, E. (1872). *Nuovo Cimento* (2), **7**, and **8**.
Bow, R. H. (1873). *Economics of Construction in Relation to Framed Structures.*
BRESSE, J. A. C. (1865). *Cours de mécanique appliquée*, Paris.
CASTIGLIANO, A. (1879). *Théorie de l'équilibre des systèmes élastiques*, Turin.
CHAN, A. S. L. (1960). College of Aeronautics Rep. No. 142.
CREMONA, L. (1872). *Le figure reciproche nella statica grafica*, Milan.
CROSS, H. (1930). *Proc. Am. Soc. Civ. Eng.* **56**, 919.
CULMANN, K. (1866). *Die graphische Statik*, Zürich.
ENGESSER, F. (1889). *Z. Architek. u. Ing. Ver. Hannover* **35**, 733.
EULER, L. (1744) *Methodus inveniendi lineas curvas. . . .*
FRÄNKEL, W. (1876). *Civiling.* **22**, 441.
GROVER, H. J., GORDON, S. A. and JACKSON, L. R. (1956). *Fatigue of Metals and Structures*, Thames & Hudson.
HEMP, W. S. (1973). *Optimum Structures.* Clarendon Press.
HENCKY, H. (1923). *Z. angew. Math. Mech.* **3**, 241.
HENNEBERG, L. (1886). *Statik der Starren Systeme*, Darmstadt.
HOFF, N. J. (1962). *Creep in Structures*, International Union on Theoretical and Applied Mechanics, Springer–Verlag.
HOOKE, R. (1678). *De Potentiâ Restitutiva.*
JOHNSON, W. and MELLOR, P. B. (1962). *Plasticity for Mechanical Engineers*, Van Nostrand.

REFERENCES

JOUKOWSKI, N. E. (1908). See *Collected Papers*, vol. 1, 1937.
JOURAWSKI, D. I. (1850). *Zhurnal Glavnago Upravienia Putej Soobchenia*, Publichnih Rabot.
LAMARLE, E. (1846). *Ann. Travaux Publics Belgique* **4**, 1.
MANDERLA, H. (1880). *Allgem. Bauztg.* **45**, 34.
MATHESON, J. A. L. (1971). *Hyperstatic Structures*. Butterworths.
MAXWELL, J. C. (1864a). *Phil. Mag.* **27**, 250.
MAXWELL, J. C. (1864b). *Phil. Mag.* **27**, 294.
MAXWELL, J. C. (1890). *Scientific Papers*, vol. 2, C.U.P.
MELAN, E. (1936). *S. B. Akad. Wiss. Wien* (Abt IIa) **145**, 195.
MÉNABRÉA, L. F. (1858). *Compt. rend.* **46**, 1056.
MICHELL, A. G. M. (1904). *Phil. Mag.* (6), **8**, No. 47, 589.
MOBIUS, A. F. (1837). *Lehrbuch der Statik*, vol. 2, Leipzig.
MOHR, O. (1868). *Z. Architek. u. Ing. Ver. Hannover* 19.
MOHR, O. (1874). *Z. Architek. u. Ing. Ver. Hannover* 223, 509.
MOHR, O. (1892). *Ziviling.* **38**, 577.
MOHR, O. (1905). *Abhandlungen aus dem Gebiete der technischen Mechanik.*
MÜLLER-BRESLAU, H. (1887a). *Schweiz. Bauztg.* **9**, 121.
MÜLLER-BRESLAU, H. (1887b). *Die graphische Statik der Baukonstruktionen.*
ODQVIST, F. K. G. (1962). *Kriechfestigkeit metallischer Werkstoffe*, Springer-Verlag.
PARKES, E. W. (1954). *Aircraft Eng.* **26**, 402.
PERRY, J. and AYRTON, W. E. (1886). *The Engineer* **57**, 464
PIPPARD, A. J. S. and BAKER, J. F. (1957). *The Analysis of Engineering Structures*, 3rd ed., Arnold.
PRAGER, W. and HODGE, P. G. (1951). *Theory of Perfectly Plastic Solids*, Wiley.
PRANDTL, L. (1923). *Z. angew. Math. Mech.* **3**, 401.
RAYLEIGH, LORD (1873). *Proc. London Math. Soc.* **4**, 357.
RITTER, A. (1862). *Elementare Theorie und Berechnung eiserner Dach und Brücken-Constructionen.*
ROBERTSON, A. (1925). *Inst. Civ. Eng. S. E. P.* **28**.
SAVIOTTI, C. (1888). *La Statica Grafica*, Milan.
SCHUR, F. (1895). *Z. Math. u. Physik* **40**, 48.
SCHWEDLER, J. W. (1851). Theorie der Brückenbalkensysteme, *Z. Bauwesen*, vol. 1, Berlin.
SOUTHWELL, R. V. (1920). *Engineering* **109**, 165.
SOUTHWELL, R. V. (1940). *Relaxation Methods in Engineering Science*, O.U.P.
THOMPSON, S. and CUTLER, R. W. (1932). *Trans. Am. Soc. Civ. Eng.* **96**, 108.
TIMOSHENKO, S. P. and GERE, J. M. (1961). *Theory of Elastic Stability*, 2nd ed., McGraw-Hill.
WHIPPLE, S. (1847). *An Essay on Bridge Building*, Utica, N.Y.
WILLIOT, M. (1877). *Notions practiques sur la statique graphique*, Paris.
WINKLER, E. (1868). *Mitte Architek. u. Ing. Ver. Böhmen* **6**.
YOUNG, THOMAS (1807). *A Course of Lectures on Natural Philosophy and the Mechanical Arts*, London.

index

algebraic solution for displacements 62, 84
alternating plasticity 157, 159
analysis 3, 17, 21, 22, 85, 129, 142, 160
arrangement of bars 47
aspect ratio 188
ASPLUND, S. O. 44

BAKER, J. F. 125, 127, 137, 149
bar 6, 27
bearing 149
bearing stress 152
BEGGS, G. E. 82
bending moment 131
bending moment–curvature relationship 131
bending of members 22, 24, 129, 130, 136
bending stiffness 132
BENDIXEN, A. 134
BETTI, E. 79
Bow's notation 34, 82
braced frames 22
BRESSE, J. A. C. 60

cantilever 31, 43, 58, 97, 188

carry over factor 139, 141
CASTIGLIANO, A. 103, 105
Castigliano's theorems 20, 105, 106, 109
CHAN, A. S. L. 188
circular fans 175, 177, 186
collapse 154
comparison of lengths 89, 100
compatibility 4, 18, 19, 21, 63, 85, 89, 93, 98, 100, 176, 186
complementary energy 20, 102, 104, 107, 108, 109, 112, 116, 117, 119, 127
compression 14, 113, 146, 162, 165, 167, 169
concentrated load 176
conservation of energy 101
constraints 121
corrosion 153
creep 6, 9, 141, 143, 149
Cremona diagram 34
critical load 147
critical time 149
CROSS, H. 127, 137
CULMANN, K. 33
curvature 131, 132, 146, 175
curvilinear coordinates 170
CUTLER, R. W. 137
cyclic loading 141, 143, 155

195

INDEX

d'Alembert's principle 3, 17
datum geometry 108, 110
dead load 18
deflection 22
deformation 2, 6, 18, 19, 141
design 3, 18, 160
direct strain 12
direct stress 12
displacement diagrams 18, 19, 55, 71, 86, 89, 96
displacements, algebraic solution 62, 84
displacements, due to loads 105
displacements, superposition of 68
displacements, total 103, 104
distribution factor 138
distribution method 127
domain of space 165, 167, 168, 180, 188
dynamic problems 3, 17

effective length 148
elastic limit 142
elastic material 9, 142
elasto-plastic material 10, 149
elongation 18, 19, 63, 107, 129, 166
energy methods 5, 19, 20, 21, 100, 110
energy of a structure 100, 106, 110
energy theorems 102, 109
ENGESSER, F. 104, 106
engineers' stress 12
environment 1, 6, 141
environment – deformation relationships 8, 19, 73, 85
equiangular spirals 175, 179
equilibrium 3, 16, 18, 19, 21, 27, 29, 45, 85, 89
Euler critical stress 148
Euler curve 147
excess of length 108, 110, 121
expansion 7, 11
experimental analysis 77, 82
extension 8, 71

fail safe 143
failure 2, 18, 129, 141
fans, see circular fans
fatigue 141, 143, 149, 153
fit 4, 19, 20, 85, 89, 93, 142

fixed-end moment 139
fixed support 30, 47
flexural rigidity 132, 137
form of optimum framework 168
FRÄNKEL, W. 50
fretting 141, 153
function 1, 17, 18, 21, 160

geometrical compatibility, see compatibility
GERE, J. M. 158
GORDON, S. A. 149
graphical methods for orthogonal nets 184
group displacement 124
group relaxation 124
GROVER, H. J. 149
gusset plate 27, 149

heat flow 101
HEMP, W. S. 188
Hencky–Prandtl nets 175
Hencky's theorem 174, 184
Henneberg's method 39, 82
HEYMAN, J. 149
HODGE, P. G. 173, 184
HOFF, N. J. 149
Hooke's law 68, 77, 82, 86, 100, 103, 105, 107, 109
HORNE, M. R. 149
humidity 1, 27, 45, 69, 101

ill-conditioned frames 24, 45, 86
initial stresses, see pre-stress
incremental collapse 141, 143, 158
influence lines 50, 81, 83, 86
instability 133, 141, 142, 148, 158
interaction of effects of environmental changes 8
internal energy 101
internal forces 18

JACKSON, L. R. 149
JOHNSON, W. 184
joints 14, 22, 23, 27, 29, 31, 45, 51, 103, 129, 137, 141, 142, 149
JOUKOWSKI, N. E. 39
JOURAWSKI, D. I. 31

INDEX

LAMARLE, E. 147
leg of weld 153
linear elastic material 9, 69, 104, 110, 136
linear system 8, 20, 50, 69, 93, 103
live load 18
load 1, 6, 18, 141, 176
load factor 142, 148

MANDERLA, H. 129
material properties 18, 19, 21, 70
MATHESON, J. A. L. 55
maximum and minimum strain 169, 170
MAXWELL, J. C. 45, 66, 89
Maxwell diagrams 34, 49, 82
Maxwell's lemma 160
Maxwell's reciprocal theorem 77, 79, 84, 86, 98, 124
mechanism 14, 22, 153
MELAN, E. 157
MELLOR, P. B. 184
MÉNABRÉA, L. F. 109
method of sections 37, 39, 82, 83
MICHELL, A. G. M. 175, 180
Michell frameworks 174, 175, 187, 190
Michell's theorem 165, 180
mild steel 9, 10
minimum weight 160
MOBIUS, A. F. 45
modulus of section 136
MOHR, O. 39, 50, 66, 89, 130
Mohr diagram 61
moment distribution 127, 137, 158
moment distribution factor 138
MÜLLER-BRESLAU, H. 39, 50

nodes 27
non-braced frames 22

OCKLESTON, A. J. 125, 127
ODQVIST, F. K. G. 149
operations, table of 123, 124

panel points 51
PARKES, E. W. 158
permanent set 8, 142
PERRY, J. 146

Perry–Robertson curve 147, 148
pin-ended member 28
PIPPARD, A. J. S. 127, 137
plastic material 10, 173, 174, 184
polygon of forces 33, 35
potential energy 104
PRAGER, W. 173, 184
Prandtl's theorem 175
pre-stress 45, 69, 70, 73, 86, 97, 157
primary stresses 130, 136, 145
proof stress 142

radius of gyration 147
RAYLEIGH, Lord 81
reactions 3, 19, 30, 45, 46, 73
reciprocal theorem, see Maxwell
rectangular net 175, 177
redundancies 17, 19, 20, 21, 87
redundant truss 16, 20, 50, 70, 73, 85, 100, 142, 155, 189
relaxation methods 20, 21, 121, 127, 137
repeated loading 141, 143, 155
residual extension 8
residual jack loads 121, 125
residual moment 139
resolution at joints 31, 34, 82, 184
restraints 19, 20
rigid-plastic material 10, 11
RITTER, A. 38, 39
rivets 27, 129, 149
ROBERTSON, A. 147
roller support 30
rotation 172

safe life 143
safety 142
SAVIOTTI, C. 39
SCHUR, F. 39
SCHWEDLER, J. W. 38
secondary stresses 129, 130, 133, 136, 146
section modulus 136
sections, method of 37, 39
settlement of foundations 142
shaft, minimum weight 180
shakedown 157
shape, change of 44, 49
shear force 131, 136
shearing of rivets 141, 149

INDEX

shear lines   173, 174
shear stress in rivet or weld   151
simply supported truss   59
singular point in strain field   177
skew-symmetry   71, 75, 93
slenderness ratio   147
slope-deflection equations   134, 141, 158
SOUTHWELL, R. V.   41, 121, 125, 137
spiral nets, *see* equiangular spirals
statical determinacy   14, 18, 21, 44, 48, 63, 85
statically determinate trusses   27, 50, 83, 100, 153, 163
statically indeterminate structure   16, 19, 21, 68, 73, 85, 153, 163
stiffness   143, 168
strain   11, 12, 71, 169, 172
strain energy   101, 103, 104, 105, 107, 108, 109, 167
stress   12, 71
stress factor   142, 145, 148
stress–strain curves   13, 137
struts   146
superposition   48, 51, 68, 77, 79, 86, 93, 97
supports   30, 45, 61, 68, 69, 113
symmetrical structure   61, 69, 75, 81, 83, 84, 86, 93
synthesis   3, 161

temperature   1, 6, 11, 27, 45, 69, 77, 83, 93, 101, 110, 117, 141
tension   14, 113, 146, 162, 165, 167, 169
tension coefficients   41, 83, 117
thermal expansion, coefficient of   11, 14

thermal strain   11
THOMPSON, S.   137
three-force system   181, 186
throat of weld   153
TIMOSHENKO, S. P.   158
torque   180
train of loads   52, 55
true direct strain   12
true direct stress   12
truss   27

uniform dilatation   178, 179
unit load method   66, 77, 93, 97
unit problem   121, 137
unloading   8

virtual deformation of space   166, 167, 170
virtual displacements   65, 119
virtual strain   166, 168, 176, 177, 189, 190
virtual work   20, 64, 77, 84, 93, 97, 103, 112, 166
volume of optimum framework   165, 167, 178, 179, 180, 183, 188

welds   27, 129, 149
WHIPPLE, S.   31
Williot diagram   56, 59, 75
Williot–Mohr diagram   61, 97
WINKLER, E.   50

yield stress   142, 147
Young's modulus   14